#4848

SCIENCE WORKSHOP SERIES

PHYSICAL SCIENCE

Matter and Energy

with an Introduction to Scientific Method

AF676864

Seymour Rosen

GLOBE BOOK COMPANY
A Division of Simon & Schuster
Englewood Cliffs, New Jersey

THE AUTHOR

Seymour Rosen received his B.A. and M.S. degrees from Brooklyn College. He taught science in the New York City School System for twenty-seven years. Mr. Rosen was also a contributing participant in a teacher-training program for the development of science curriculum for the New York City Board of Education.

Cover Photograph: Adam Hart-Davis/Science Photo Library
Photo Researcher: Rhoda Sidney

Photo Credits:

p. 21, Fig. A: Grant Heilman
p. 21, Fig. B: Strickler/Monkmeyer Press
p. 21, Fig. C: UPI/BETTMANN
p. 21, Fig. D: United Nations
p. 24, Fig. E: Honda
p. 24, Fig. F: Ford Motor Company
p. 24, Fig. G: UPI
p. 24, Fig. H: American Airlines
p. 30, Fig. E: Ward's Natural Science
p. 30, Fig. F: Rhoda Sidney
p. 35, Fig. E: Rhoda Sidney
p. 35, Fig. F: Rhoda Sidney
p. 42, Fig. E: Rhoda Sidney
p. 42, Fig. F: Rhoda Sidney
P. 42, Fig. G: Rhoda Sidney
p. 45, Fig. A: Spencer Grant/Monkmeyer Press
p. 45, Fig. B: George Zimmer/Monkmeyer Press
p. 47, Fig. E: NASA
p. 51, Fig. A: Runk/Schoenberger, Grant Heilman
p. 51, Fig. B: Runk/Schoenberger, Grant Heilman
p. 51, Fig. C: New York Convention and Visitor Center
p. 53, Fig. F: Dan McCoy/Rainbow
p. 89, Fig. G: United Nations
p. 89, Fig. H: American Airlines
p. 89, Fig. I: Helena Frost
p. 89, Fig. J: Barry L. Runk/Grant Heilman
p. 100, Fig. E: Michael Weisbrot/Stock, Boston
p. 100, Fig. G: Courtesy of British Government
p. 118, Fig. F: NASA
p. 120, Fig. G: NASA
p. 139, Fig. C: James Carroll
p. 139, Fig. D: PhotoResearchers
p. 160, Fig. O: Swiss National Tourist Office

ISBN: 0-8359-0278-1

Printed in the United States of America
3 4 5 6 7 8 9 10 95 94 93

GLOBE BOOK COMPANY
A Division of Simon & Schuster
Englewood Cliffs, New Jersey

CONTENTS

SCIENTIFIC METHODS

STATES OF MATTER

ATOMS AND ELEMENTS

MOTION

SIMPLE MACHINES

Introduction to Physical Science

The ancient Egyptians built huge pyramids to protect the bodies of their kings. The largest pyramid—the Great Pyramid—was built about 4500 years ago. The Great Pyramid is 450 meters high. It is made up of two million stone blocks. Each block weighs approximately 2 1/2 metric tons.

How were the ancient Egyptians able to build this marvel of ancient architecture? Afterall, they had no modern machinery or tools. The ancient Egyptians used ramps to build the pyramid. After building the first layer, workers built long ramps of dirt and brick. Then they dragged the stones up the ramp to make the next layer. As each layer was finished, the ramps were raised and lengthened.

In this book, you will learn how ramps and other simple machines make work easier to do. However, the study of simple machines is just one part of physical science. You also will learn about matter and energy, and how they are related. You will learn about atoms—the building blocks of all matter—and the different forms of energy. But most important, after completing this book, you will have a better understanding of how the physical world works.

How do scientists measure things?

1

area: measure of the size of a surface
length: distance between two points
mass: amount of matter in an object
temperature: measure of how hot or cold something is
volume: measure of the amount of space an object takes up
weight: measure of the pull of gravity on an object

LESSON 1 | How do scientists measure things?

How much do you weigh? What is your height? How many tiles will cover your kitchen floor? How much milk should be added to a cake mix? What is the temperature outside? All of these questions are answered by measurements.

Measuring is an important part of daily life. People use measurements all the time—for shopping, cooking, construction, and deciding how warm to dress. Measuring also is an important part of science.

A measurement has two parts: a number and a unit. A unit is a standard amount used to measure something.

EXAMPLES

100	grams	25	liters
number	standard unit	number	standard unit

There are many kinds of measurements. The most common are:

MASS and weight are related, but they are not the same. **Mass** is a measure of the amount of matter in an object. **Weight** is a measure of the pull of gravity on an object. The basic unit of mass in the metric system is the kilogram (kg). Mass is measured with a balance.

LENGTH is the distance from one point to another as measured by a ruler. The basic metric unit of length is the meter (m). You can use a meter stick or metric ruler to measure length.

AREA is a measure of surface room—how big something is in two directions. You can find the area of a rectangle by multiplying its length by its width. Area is measured in square units, such as square meters (m^2).

VOLUME is the measure of the amount of space an object takes up—how big an object is in all three directions. The liter (L) is the basic unit of volume in the metric system. A measuring cup or a graduated cylinder is used to measure the volume of liquids.

The volume of a solid can be measured in cubic centimeters (cm^3). You can find the volume of a cube or rectangle by multiplying its length by its width by its height. 1000 cubic centimeters equals 1 liter.

TEMPERATURE is the measure of how hot or cold an object is. Temperature is measured with a thermometer in degrees Celsius °C, or degrees Fahrenheit °F. The Celsius scale usually is used in science.

UNDERSTANDING METRICS

In the United States, people usually use English units of measurement such as, ounces, pounds, inches and feet. Most other countries use metric units. Metric units include the gram, kilogram, meter, and centimeter. Scientists also use the metric system. In science, you will use mostly metric units.

The metric system is based upon units of ten. Each unit is ten times smaller or larger than the next unit. This means that a unit is made larger by multiplying it by 10 and made smaller by dividing by 10. Prefixes describe a unit's value. The prefixes and their meanings are listed below.

PREFIX	MEANING	
kilo- [KILL-uh]	one thousand (1,000)	each, larger by a multiple of ten
hecto- [HEC-tuh]	one hundred (100)	
deca- [DEC-uh]	ten (10)	
deci- [DESS-ih]	one tenth (1/10)	each, smaller by a multiple of 1/10
centi- [SEN-tih]	one hundredth (1/100)	
milli- [MILL-ih]	one thousandth (1/1,000)	

Use the chart above to answer the following questions.

1. To change from tens to hundreds, you multiply by ______________ .
   1, 10, 100

2. To change from hundreds to thousands, you multiply by ______________ .
   1, 10, 100

3. In the metric system, to change from one unit to the next higher unit, what must you do?

   ____

4. To change from one unit to the next lower unit, you must divide by ______________ .
   1, 10, 100

5. Which prefix stands for a greater value?

   a) deca- or kilo-? ______________
   b) kilo- or milli-? ______________
   c) centi- or milli-? ______________
   d) hecto- or kilo-? ______________
   e) centi- or deci-? ______________
   f) deca- or deci-? ______________

MEASURING MASS

1. In the metric system, the unit of <u>mass</u> is the ________________ .
meter, kilogram, pound

2. Mass and weight ________________ the same.
are, are not

3. ________________ is a measure of the amount of matter in an object.
mass, weight

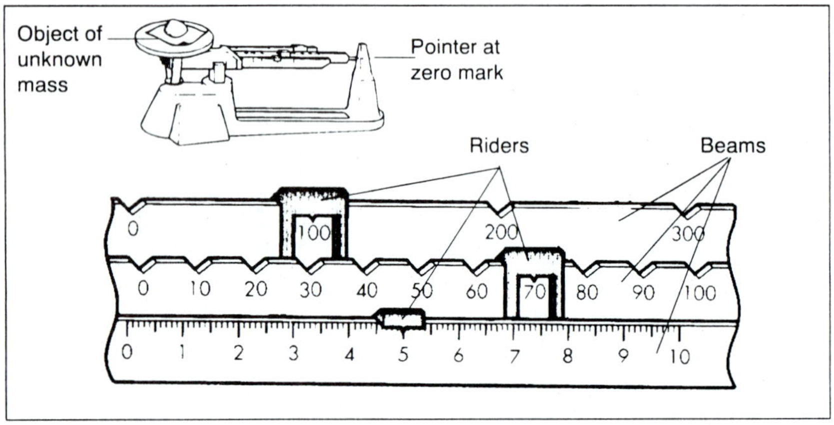

Figure A

4. What instrument is used to measure mass? ________________

5. What is the mass of the object shown? ________________

TRUE OR FALSE

In the space provided, write "true" if the sentence is true. Write "false" if the sentence is false.

________ **1.** Weight is a measure of the pull of gravity on an object.

________ **2.** Scientists use English units of measurement.

________ **3.** The prefix centi- stands for one hundredth (1/100).

________ **4.** A graduate is used to measure mass.

________ **5.** The basic unit of length in the metric system is the meter.

________ **6.** Volume is a measure of the amount of matter in an object.

________ **7.** One kilogram is less than one gram.

________ **8.** A measurement has two parts.

________ **9.** A unit is an amount used to measure something.

________ **10.** Most countries use the metric system.

MEASURING LENGTH

Length is measured with a metric ruler. Part of a combined metric and inch ruler is shown in Figure B. On the metric side of the ruler the distance between numbered lines is equal to one centimeter. Each centimeter is divided into 10 equal parts. Each one of these parts is equal to one millimeter.

The figure below shows a combined metric and inch ruler.

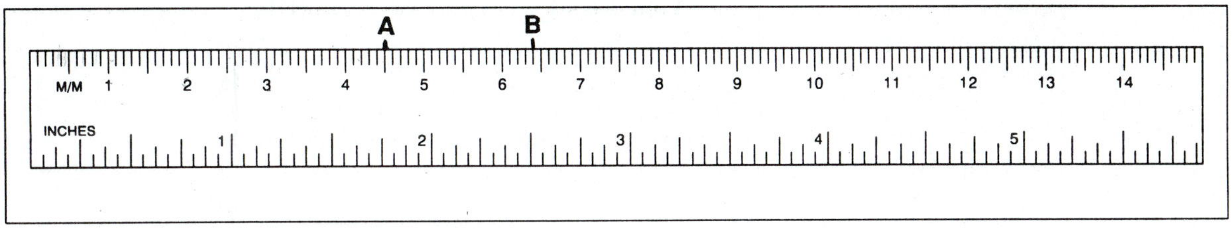

Figure B

1. What value does the prefix <u>milli-</u> stand for? ____________________

2. What value does the prefix <u>centi-</u> stand for? ____________________

3. Which is <u>larger</u>, a meter or a millimeter? ______________

4. How many millimeters make 1 centimeter? ______________

5. The length at A may be written as 45 mm. It may also be written as ______________.
45 cm, 4.5 cm, 4.5 mm

6. The length at B may be written as ______________ mm or ______________ cm.

Measure each of the following lengths. Write the lengths on the right in centimeters and millimeters.

7. ____________________________________ **7.** _____ cm _____ mm

8. _________________________ **8.** _____ cm _____ mm

9. ________ **9.** _____ cm _____ mm

10. ____ **10.** _____ cm _____ mm

To the right of each length listed, <u>draw</u> a line of that length.

a) 92 mm

b) 9.2 cm

c) 43 mm

d) 3.5 cm

MEASURING AREA

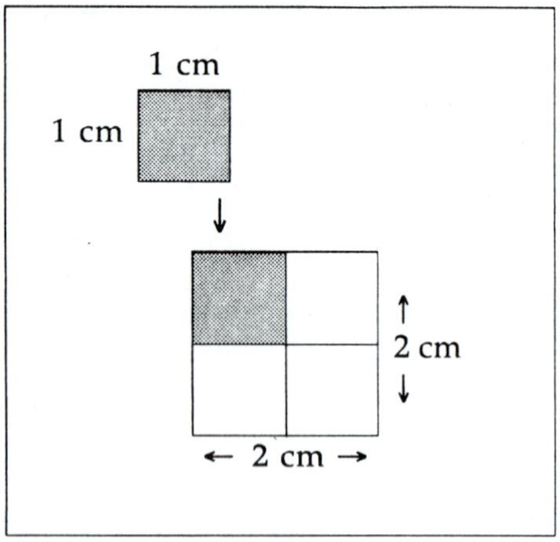

The square in Figure C has an area of 2 square centimeters (2 cm^2).

Area = l × l_2

= 2 cm × 2 cm

Area = 4 square centimeters (4cm^2).

Figure C

Figure the area of each of the following rectangles: (Measure Figures G and H yourself.)

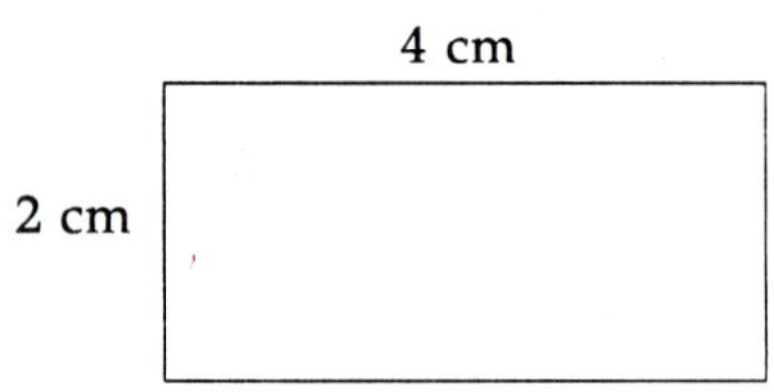

Figure D

____

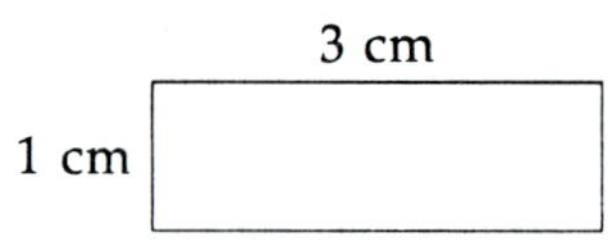

Figure E

____

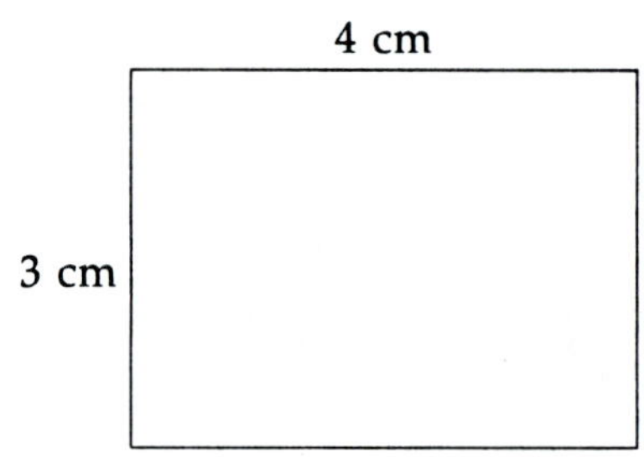

Figure F

____

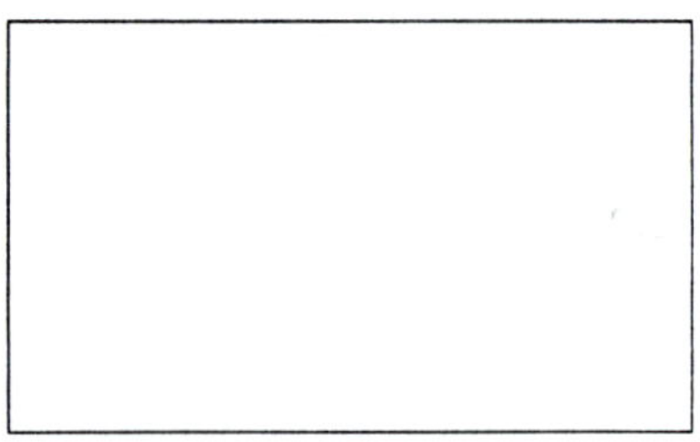

Figure G

____

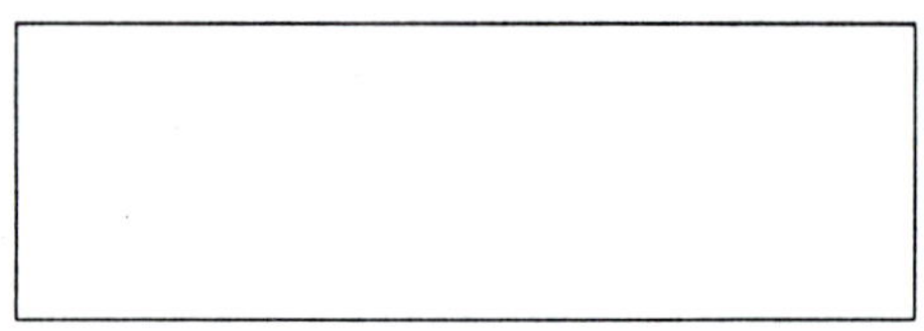

Figure H

____

CALCULATING AREA

Find the areas of the following rectangles:

1. 5 meters × 5 meters ________________
2. 2.5 cm × 5 cm ________________
3. 10 millimeters × 10 millimeters ________________

MEASURING VOLUME

The volume of liquids is measured in a graduated cylinder. A graduated cylinder is a glass tube that is marked with divisions to show the amount of liquid in it. To measure liquid volume, you hold the graduated cylinder at your eye level. The surface of the liquid will have a "belly-down" curve. You should read the mark that lines up with the bottom of the curve.

What is the liquid volume in this graduated cylinder?

____

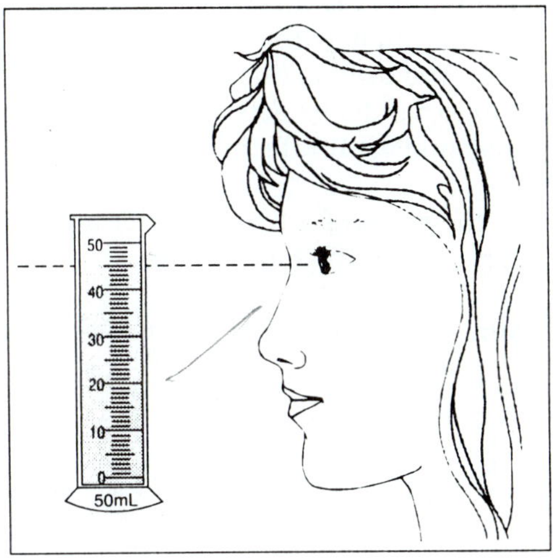

Figure I

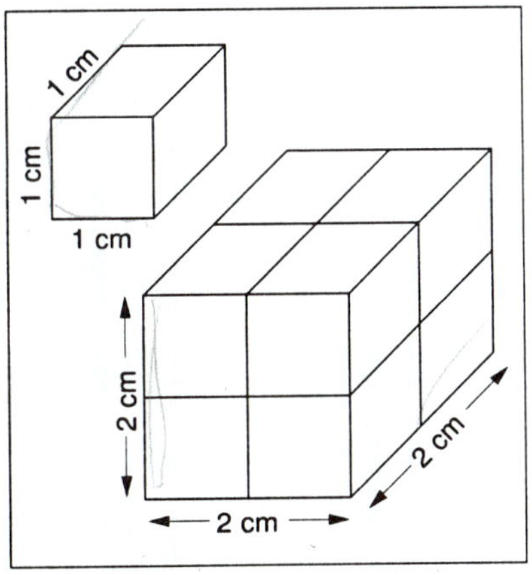

Figure J

What is the volume of a cube that is 2 cm × 2 cm × 2 cm?

Volume $= l_1 \times l_2 \times l_3$

$= 2$ cm × 2 cm × 2 cm

Volume = 8 cubic centimeters ($8 cm^3$).

Find the volume of each of the following rectangles:

	Volume
1. 2 cm × 5 cm × 1 cm	________________
2. 8 m × 2 m × 2 m	________________
3. 1 mm × 1 mm × 10 mm	________________
4. 4 cm × 2 cm × 3 cm	________________
5. 5 m × 3 m × 6 m	________________

READING A CELSIUS THERMOMETER

Temperature is measured with a thermometer. Many thermometers, including the ones you are most familiar with, are made of glass tubes. At the bottom of the tube is a wider part called the bulb. The bulb is filled with a liquid, such as mercury. When the bulb is heated, the liquid in the bulb expands, or gets larger. It rises in the tube. When the bulb is cooled, the liquid contracts, or gets smaller. It falls in the tube.

On the sides of a thermometer are a series of marks. You read the temperature by looking at the mark where the liquid stops.

Write the temperature shown on each celsius thermometer in the space provided.

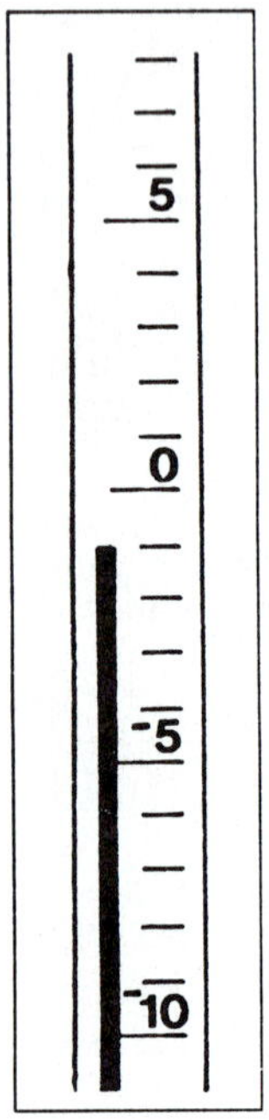

Figure K ____________

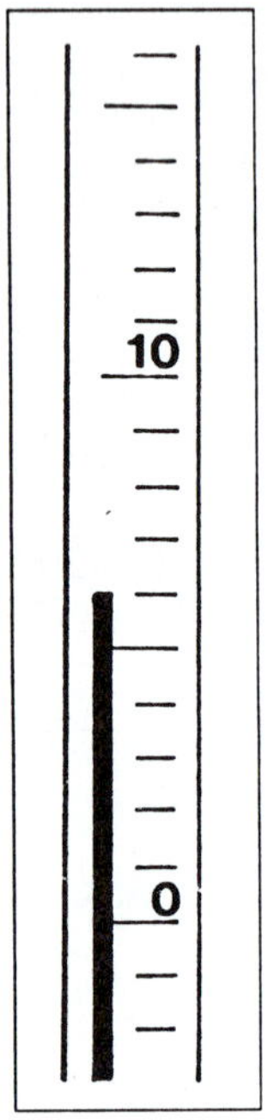

Figure L ____________

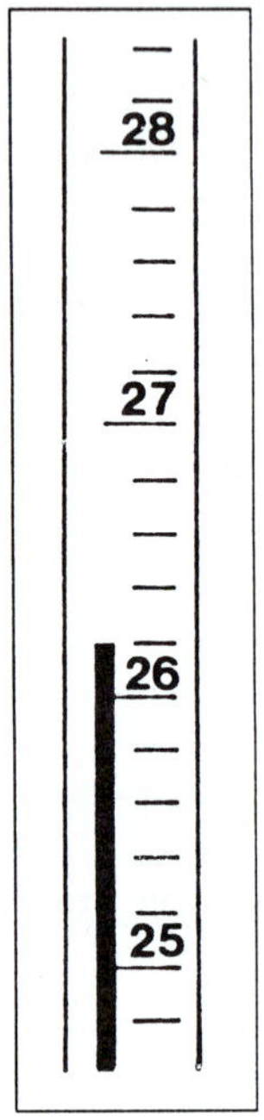

Figure M ____________

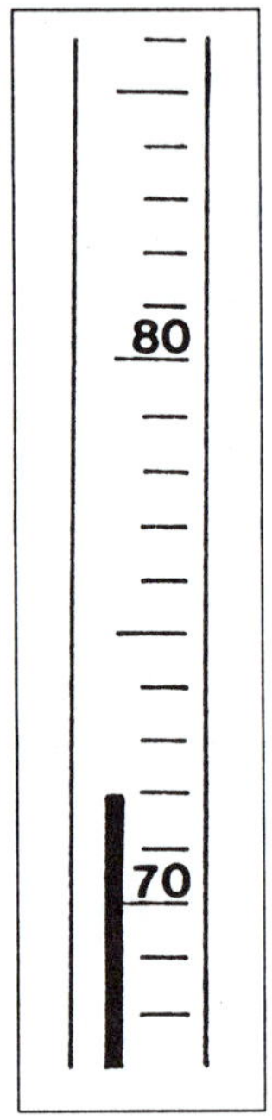

Figure N ____________

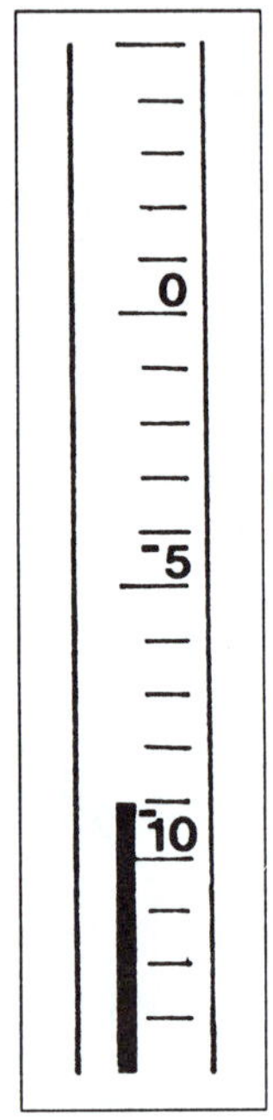

Figure O ____________

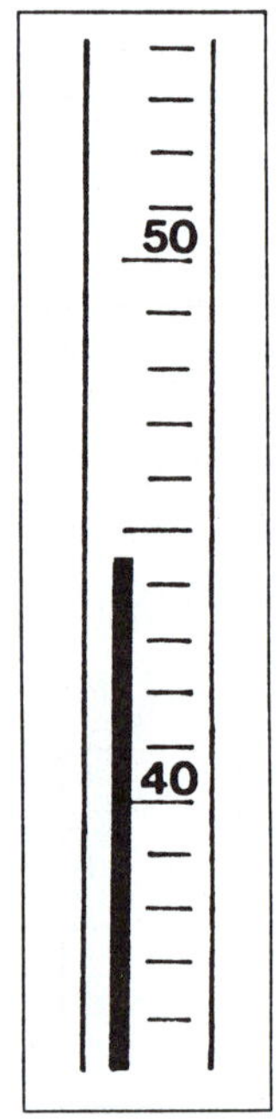

Figure P ____________

One cubic centimeter is equal to one milliliter (mL). How many <u>liters</u> of water can a 1800 cm^3 pan hold? ______________

What is scientific method?

2

data [DAYT-uh]: record of observations
hypothesis [hy-PAHTH-uh-sis]: suggested solution to a problem based upon known information
scientific method: problem solving guide

LESSON 2 | What is scientific method?

You may not realize it, but you do problem-solving every day. You do not always think about how to solve a particular problem. You solve problems in sort of a "natural" way, a way that seems to "make sense." And it usually does.

For example, suppose you put your key into your house door and try to turn it, but it does not budge. You wonder what's wrong. You examine the key to make certain that it is the correct one. Then you try again. The key still does not turn. What next? You might "jiggle" the key. Or, you might pull back on the door knob as you try to turn the key. One of these approaches <u>might</u> work. If not, you try other methods until the problem is solved.

Without knowing it, you solve problems very much like a scientist does. You use **scientific method**. Scientific method is a guide used to solve problems. It involves asking questions, making observations, and trying things out in an orderly way. Scientists use certain steps to solve problems. The steps of scientific method are:

- **IDENTIFY THE PROBLEM** State it clearly—usually as a question.
- **GATHER INFORMATION** Research; ask questions. Discover what is already known about the problem.
- **STATE A HYPOTHESIS** A **hypothesis** [hy-PAHTH-uh-sis] is a suggested solution as to why something happens.
- **TEST THE HYPOTHESIS** Experiment and examine the situation to check the hypothesis.
- **MAKE CAREFUL OBSERVATIONS** Note everything your senses can gather. Record the **data** [DAYT-uh]. Keep careful records.
- **ORGANIZE AND ANALYZE THE DATA** Put the data in order. Scientists often use charts and tables to organize data. Figure out the <u>meaning</u> of the data.
- **STATE A CONCLUSION** Explain the data. State whether or not it supports the hypothesis.

Different problems require different approaches. Not every step in scientific method needs to be used. And the steps can be used in any order.

CHOOSE THE RIGHT CAPTION

Below are six figures and six captions. Each caption matches one of the figures. Choose the caption that best describes each figure. Write the correct caption on the line provided.

Choose from these captions:

Identify the problem	Make careful observations
Gather information	Record the data
State a hypothesis	Analyze the data
Test the hypothesis	State a conclusion

Figure A

1. ______________________________

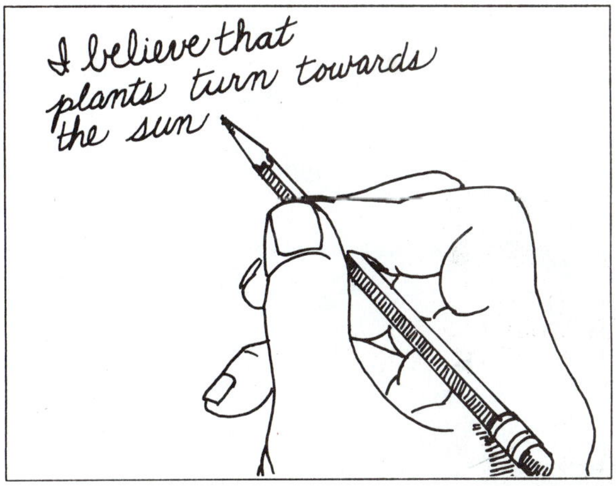

Figure B

2. ______________________________

Figure C

3. ______________________________

Figure D

4. ______________________________

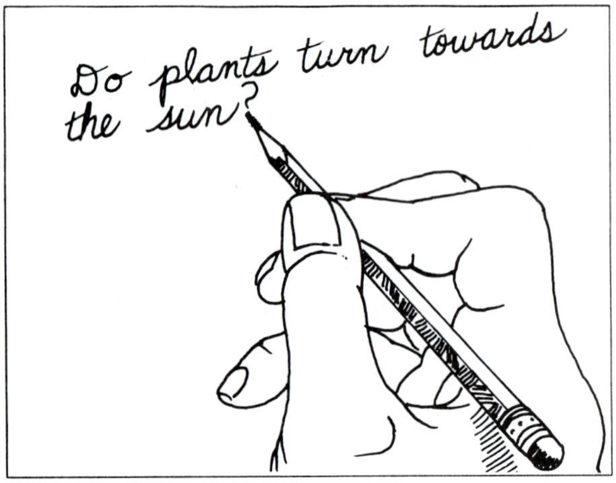

Figure E

5. ______________________

Figure F

6. ______________________

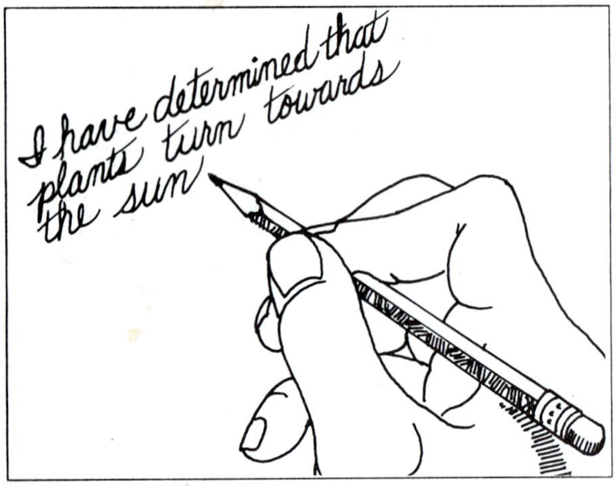

Figure G

7. ______________________

Figure H

8. ______________________

MATCHING

Match each term in Column A with its description in Column B. Write the correct letter in the space provided.

	Column A	Column B
______	**1.** analyze	**a)** explains the data
______	**2.** scientific method	**b)** suggested solution
______	**3.** conclusion	**c)** test the hypothesis
______	**4.** hypothesis	**d)** guide for solving problems
______	**5.** experiment	**e)** figure out the meaning

USING SCIENTIFIC METHOD

Two separate stories are shown in the figures below. However, the figures in each are not in the proper order (sequence). In the table under each set of figures, list the figures in their proper order. Also, explain what is happening in each figure. Finally, write a hypothesis (in question form) and a conclusion.

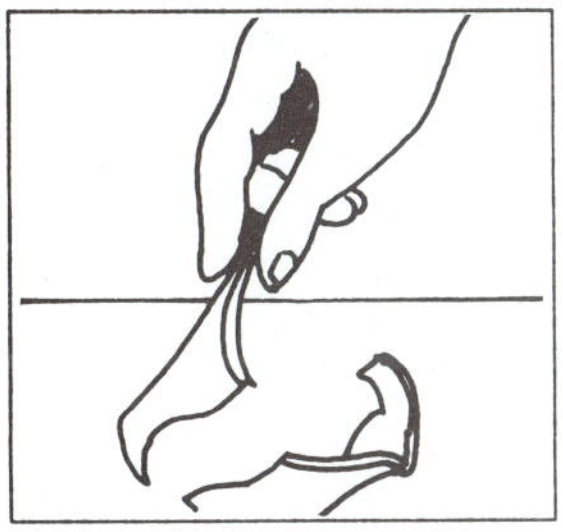

Figure I

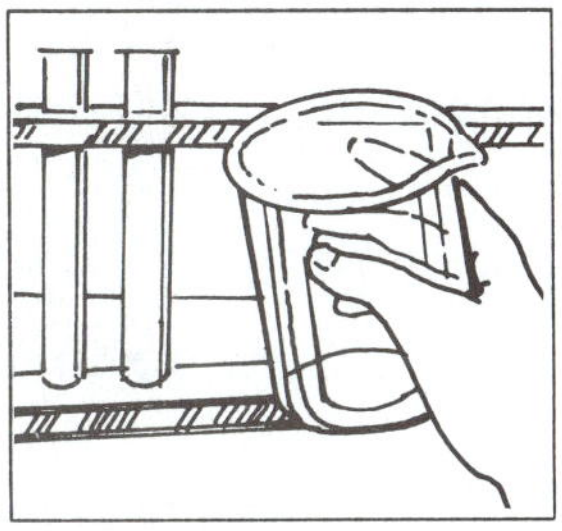

Figure J

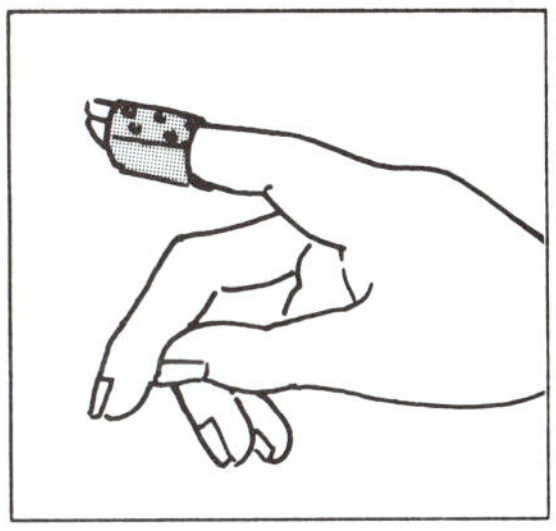

Figure K

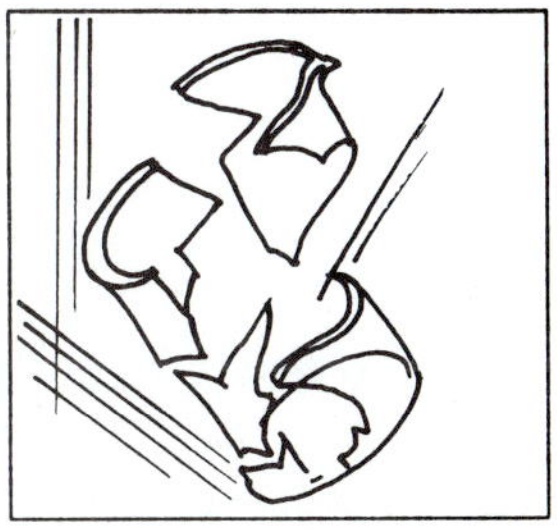

Figure L

Step	Figure	Explanation
1.		
2.		
3.		
4.		

Hypothesis: ______________________________

Conclusion: ______________________________

Figure M

Figure N

Figure O

Figure P

Step	Figure	Explanation
1.		
2.		
3.		
4.		

Hypothesis: ______________________________

Conclusion: ______________________________

FILL IN THE BLANK

Complete each statement using a term or terms from the list below. Write your answers in the spaces provided.

supports	observe	different
problems	question	already known
data	scientific method	senses
steps		

1. To test a hypothesis, scientists may ________________ natural events.
2. When scientists research, they may find out what is ________________ about a problem.
3. Your ________________ gather information.
4. A conclusion states whether or not data ________________ a hypothesis.
5. A problem is usually stated as a ________________ .
6. Scientists use certain ________________ to solve problems.
7. You solve ________________ much like scientists do.
8. Different problems can be solved in ________________ ways.
9. A guide used to solve problems is called ________________ .
10. Scientists use charts to put ________________ in order.

REACHING OUT

Jennifer has never eaten asparagus. She is afraid that it might make her sick. At dinner, she eats some. She likes the taste, but soon she suffers from nausea. Jennifer concludes that asparagus makes her sick.

1. Why might Jennifer's conclusion be incorrect? ________________________________

____

____

2. What might be done to further test her conclusion? ________________________________

____

____

How are experiments done safely?

3

caustic: able to burn and irritate the skin
safety alert symbols: signs that warn of hazards or dangers

LESSON 3 | How are experiments done safely?

"Hands-on" experiences are part of many school activities. Science, especially, is suited to "learning by doing." You investigate; you make things happen; you learn from what you do.

Science investigations can be exciting. However, they can also be dangerous. Science laboratories have equipment and materials that can be dangerous if not handled properly. For this reason, it is important for you to always follow proper safety guidelines. Safety rules are for your own protection-as well as the protection of everyone around you.

The safety rules that you should follow are listed on page 17. Read over these safety rules carefully. Notice the **safety alert symbols** that accompany the safety rules. In this book, safety alert symbols are included at the beginning of some activities to make you aware of safety precautions. Always note any safety symbols and caution statements in an activity.

To avoid accidents in the science laboratory, always follow your teacher's directions. You should not perform activities without directions from your teacher. You also should never work in the science laboratory alone.

One hazard has no symbol even though it probably causes more accidents than any others. That hazard is "horsing around." Horsing around in the laboratory can lend to serious injury—or even death. So THINK before doing anything foolish.

SAFETY ALERT SYMBOLS

CLOTHING PROTECTION • A lab coat protects clothing from stains. • Always confine loose clothing.

EYE SAFETY • Always wear safety goggles. • If anything gets in your eyes, flush them with plenty of water. • Be sure you know how to use the emergency wash system in the laboratory.

FIRE SAFETY • Never get closer to an open flame than is necessary. • Never reach across an open flame. • Confine loose clothing. • Tie back loose hair. • Know the location of the fire-extinguisher and fire blanket. • Turn off gas valves when not in use. • Use proper procedures when lighting any burner.

POISON • Never touch, taste, or smell any unknown substance. Wait for your teacher's instruction.

CAUSTIC SUBSTANCES • Some chemicals can irritate and burn the skin. If a chemical spills on your skin, flush it with plenty of water. Notify your teacher without delay.

HEATING SAFETY • Handle hot objects with tongs or insulated gloves. • Put hot objects on a special lab surface or on a heat-resistant pad; never directly on a desk or table top.

SHARP OBJECTS • Handle sharp objects carefully. Never point a sharp object at yourself-or anyone else. • Cut in the direction away from your body.

TOXIC VAPORS • Some vapors (gases) can injure the skin, eyes, and lungs. Never inhale vapors directly. Use your hand to "wave" a small amount of vapor towards your nose.

GLASSWARE SAFETY • Never use broken or chipped glassware. • Never pick up broken glass with your bare hands.

CLEAN UP • Wash your hands thoroughly after any laboratory activity.

ELECTRICAL SAFETY • Never use an electrical appliance near water or on a wet surface. • Do not use wires if the wire covering seems worn. • Never handle electrical equipment with wet hands.

DISPOSAL • Discard all materials properly according to your teacher's directions.

PUTTING SAFETY RULES TO USE

Answer the following questions in complete sentences.

1. Jean has long hair. What should she do before working near an open flame. ____________

____

2. A glass tube has broken. How should you pick up the pieces. ____________

____

3. Why should you always wear safety goggles during <u>every</u> lab activity? ____________

____

4. What else should you wear? Why? ____________

____

5. A chemical spills on your hand. You are pretty sure that it is harmless. But you are not certain.

   What should you do? ____________

IDENTIFYING SAFETY ALERT SYMBOLS

Six safety alert symbols are shown below. Match them with their meanings. Write the correct <u>letter</u> next to each description.

a.	**b.**	**c.**	**d.**	**e.**	**f.**

________ **1.** electrical safety

________ **2.** fire safety

________ **3.** heating safety

________ **4.** clothing protection

________ **5.** sharp objects

________ **6.** glassware safety

REACHING OUT

In the box at the right, design a NO HORSING AROUND symbol. Either draw it or describe it, or both. Perhaps you can think up more than one.

Is it matter or energy?

4

characteristic [kar-ik-tuh-RIS-tik]: distinguishing trait, feature, or quality
energy: ability to do work; ability to make things move
matter: anything that has mass and takes up space
property [PROP-ur-tee]: characteristic; used to describe a substance

LESSON 4 | Is it matter or energy?

Think of the world around you . . . Everywhere there are things — **matter**. You see, hear, smell, taste, and feel them. The world has many kinds of matter. There are water and rocks; wood and metal; plants and animals. Even the invisible air you breathe is matter. The list can go on and on.

The world also has **energy** — like light, heat, sound, motion, and electricity.

The entire universe is made up of matter and energy. THERE IS NOTHING ELSE.

What are the differences between matter and energy?

MATTER
- Matter has mass.
- Matter takes up space.

Therefore, matter is anything that has mass and takes up space.

ENERGY
- Energy is not like matter.
- Energy does not have mass.
- Energy does not take up space.
- Energy MOVES matter!

Therefore, energy is the ability to make things move.

Light, heat, sound, motion, and electricity, are forms of energy. Why? Because they can make things move.

In the next few lessons you will study matter. You will learn about the **properties** (PROP-ur-teez) of matter.

What are properties?

Properties or characteristics (kar-ik-tuh-RIS-tiks) are the special things that help us identify matter. You think about matter all the time. Is it red or is it blue. Is it heavy or is it light? What does it smell like?

Color, mass, and odor are three properties. There are many more.

Every kind of matter has its own properties. Some kinds of matter share properties. However, no two kinds of matter have all the same properties.

WHICH IS MATTER? WHICH IS ENERGY?

Each picture shows matter and energy. Figure out what is matter in each picture. You cannot see energy, but you can see what it does. What kinds of energy are shown in each picture? Write your answers on the lines under each picture.

Figure A

Matter ____________________

Energy ____________________

Figure B

Matter ____________________

Energy ____________________

Figure C

Matter ____________________

Energy ____________________

Figure D

Matter ____________________

Energy ____________________

FILL IN THE BLANK

Complete each statement using a term or terms from the list below. Write your answers in the spaces provided. Some answers may be used more than once.

electricity	light	mass
move	characteristics	energy
heat	smell	color
properties	matter	motion

1. Everything in the world is either ________________ or ________________ .
2. ________________ has no mass.
3. ________________ does not take up space.
4. Energy can make something ________________ .
5. Four kinds of energy are ________________ , ________________ , ________________ , and ________________ .
6. ________________ takes up space.
7. ________________ has mass.
8. Another word for characteristics is ________________ .
9. Three properties of matter are ________________ , ________________ , and ________________ .
10. Another word for properties is ________________ .

MATCHING

Match each term in Column A with its description in Column B. Write the correct letter in the space provided.

	Column A	Column B
________	1. color	a) does not take up space and has no mass
________	2. matter	b) kind of energy
________	3. energy	c) a property of matter
________	4. properties	d) takes up space and has mass
________	5. heat	e) help us identify matter

TRUE OR FALSE

In the space provided, write "true" if the sentence is true. Write "false" if the sentence is false.

_________ **1.** Matter and energy are the same.

_________ **2.** Energy has mass.

_________ **3.** Energy takes up space.

_________ **4.** Matter has mass.

_________ **5.** Matter takes up space.

_________ **6.** Iron and air are examples of matter.

_________ **7.** Light and sound are examples of matter.

_________ **8.** Properties help us identify matter.

_________ **9.** Odor is of a property of matter.

_________ **10.** Two things are tested. Their properties are found to be exactly the same. This proves they are different.

IS IT MATTER OR ENERGY?

Tell whether each of the following is matter or energy. Write "M" for matter and "E" for energy on the line provided.

_________ **1.** Has mass and takes up space.

_________ **2.** Sometimes cannot be seen but has mass and takes up space.

_________ **3.** Has no mass. Takes up no space.

_________ **4.** Has no mass. Takes up no space, but is able to move something.

_________ **5.** a light bulb

_________ **6.** the light from a light bulb

_________ **7.** a piece of coal

_________ **8.** the heat from a burning piece of coal

_________ **9.** the sound of a bugle

_________ **10.** a bugle

COMPARING CHARACTERISTICS

The two objects in each pair below share some characteristics. However, some of their characteristics are different. In the spaces provided, list the characteristics that are the same and the characteristics that are different.

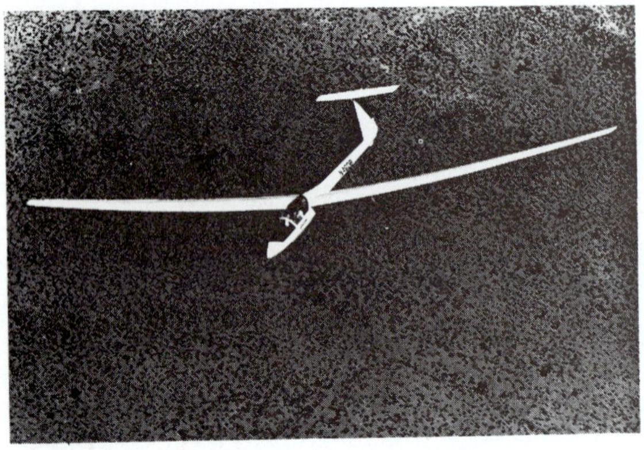

Figure E

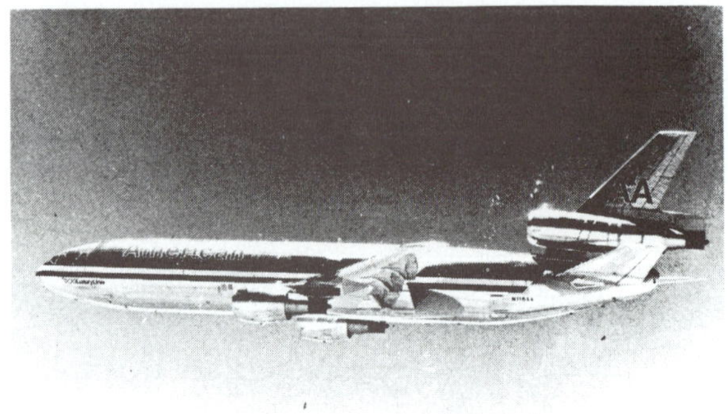

Figure F

Same Characteristics ______________________________

Different Characteristics ______________________________

Figure G

Figure H

Same Characteristics ______________________________

Different Characteristics ______________________________

REACHING OUT

A friend tells you one characteristic about an object. Why would it be hard for you to guess what the object is?

____

____

What are solids?

5

solid: phase of matter with a definite shape and a definite volume
states: phases of matter

LESSON 5 | What are solids?

Matter can be a solid, a liquid, a gas, or a plasma. Solid, liquid, gas, and plasma are the four **states**, or phases, of matter. Plasma is <u>very</u> rare on Earth but common in space. It is very hot. You probably have some idea of what we mean by solid, liquid, and gas. You probably know that a baseball is solid, water is a liquid, and air is a gas.

Sometimes, it is difficult to tell if matter is solid or liquid, or liquid or gas.

All solids have certain properties. The properties that are the same for all solids help us to decide whether something is a solid. In addition, they help us make rules that define solids as a group.

Solids are like all matter. They take up space. They have mass.

A solid takes up <u>a definite amount of space</u>. It does not keep changing the amount of space it takes up.

A solid also has <u>a definite shape</u>. It does not keep changing its shape. When a solid is in a container, it does not change its shape. It does not take the shape of the container.

Solids come in many colors, masses, shapes, and hardnesses. These properties help us tell one solid from another.

WHAT DO THE PICTURES SHOW?

Each of these figures shows at least one property of solids. Look at each figure. Then answer the questions. Circle the letter of the figure that best answers each question. You may use a figure more than once.

Figure A

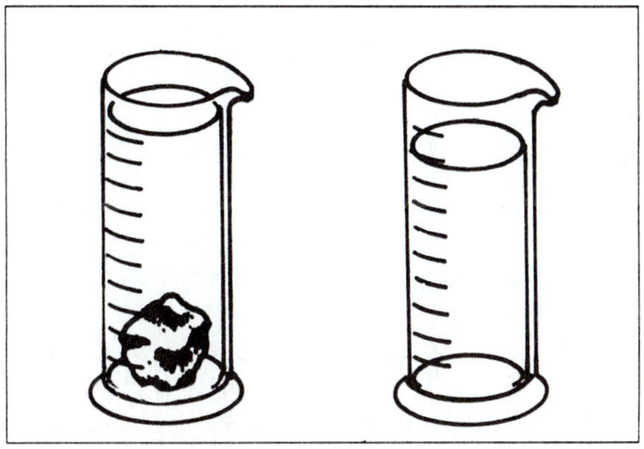

Figure B

Figure C

Figure D

Which figure shows that a solid . . .

1. . . . takes up space? A B C D

2. . . . does not take the shape of its container? A B C D

3. . . . does not fill a container larger than itself? A B C D

4. . . . has mass? A B C D

5. . . . has a definite shape? A B C D

FILL IN THE BLANK

Complete each statement using a term or terms from the list below. Write your answers in the spaces provided. Some answers may be used more than once.

solid	space	gas
masses	four	hardnesses
definite	colors	liquid
mass	plasma	rare

1. There are ______________ states of matter.
2. The four states of matter are: ______________, ______________, ______________, and ______________.
3. All matter has ______________ and takes up ______________.
4. A solid takes up a ______________ amount of space.
5. Solids come in different ______________, ______________, and ______________.
6. Plasma is very ______________ on Earth.
7. Solids are very ______________ on Earth.

TRUE OR FALSE

In the space provided, write "true" if the sentence is true. Write "false" if the sentence is false.

__________ 1. A solid is the only form of matter.

__________ 2. Every solid has mass.

__________ 3. A solid takes up a definite amount of space.

__________ 4. A mountain is an example of a solid.

__________ 5. A piece of dust is an example of a solid.

__________ 6. A solid shapes itself like a container.

__________ 7. A small amount of a solid fills a large container.

__________ 8. Every solid of the same size has the same mass.

__________ 9. Solids come in different colors.

__________ 10. Every solid has an odor.

MATCHING

Match each term in Column A with its description in Column B. Write the correct letter in the space provided.

	Column A	Column B
________	**1.** energy	**a)** the states of matter
________	**2.** hardness	**b)** has mass and takes up space
________	**3.** properties	**c)** has no mass and takes up no space
________	**4.** matter	**d)** a property of solids
________	**5.** solid, liquid, gas, plasma	**e)** help to identify matter

COMPLETING SENTENCES

Choose the correct word or term for each statement. Write your choice in the spaces provided.

1. A very small solid ________________ have mass.
does, does not

2. A piece of dust ________________ take up space.
does, does not

3. A solid ________________ take on the shape of its container.
does, does not

4. A solid ________________ spread to fill its container.
does, does not

5. Solids come in different ________________ .
states, hardnesses

WORD SCRAMBLE

Below are several scrambled words you have used in this Lesson. Unscramble the words and write your answers in the spaces provided.

1. DOLSI ________________

2. PHEAS ________________

3. RATTEM ________________

4. MASALP ________________

5. SMAS ________________

Figure E

COMPLETE THE CHART

Answer the questions by putting a YES or NO in the space provided.

		Have Mass	Take Up Space	Take Up A Definite Amount Of Space	Have Definite Shapes	Spread Outward To Fill Their Containers	Small Amounts Fill Large Containers
1.	solids						

REACHING OUT

When sand is placed in a container it looks as if it takes on the exact shape of the container. It really does not. How can you show that is does not?

Figure F

What are liquids?

6

liquid: phase of matter with a definite volume, but no definite shape

LESSON 6 What are liquids?

Water is probably the liquid that you know best. For living things, it is the most important liquid. Living things could not survive without water.

There are many other kinds of liquids. All **liquids** have certain properties that are the same.

A liquid is like all other matter. Every liquid has mass, but liquids can have different masses. Mercury is a very heavy liquid. It is about thirteen times heavier than water.

All liquids take up space. They take up a definite amount of space. We can measure how much space a liquid takes up.

Liquids have no definite shape. They flow and spread out to take the shape of their containers. If a liquid is not moving it will always be flat on top.

Solids have a definite shape. Liquids do not have a definite shape. This is an important difference between liquids and solids.

Liquids come in different colors and masses. These properties help us tell one liquid from another.

WHAT DO THE PICTURES SHOW?

Each of these figures shows at least one property of liquids. Look at each figure. The answer the question. Circle the letter of the figure that best answers each question. You may use a figure more than once.

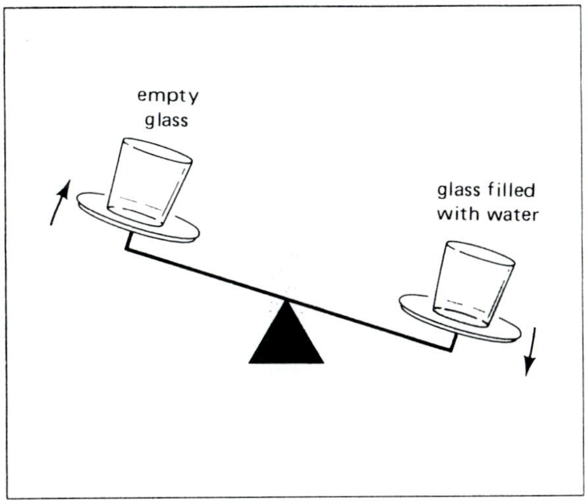

Figure A

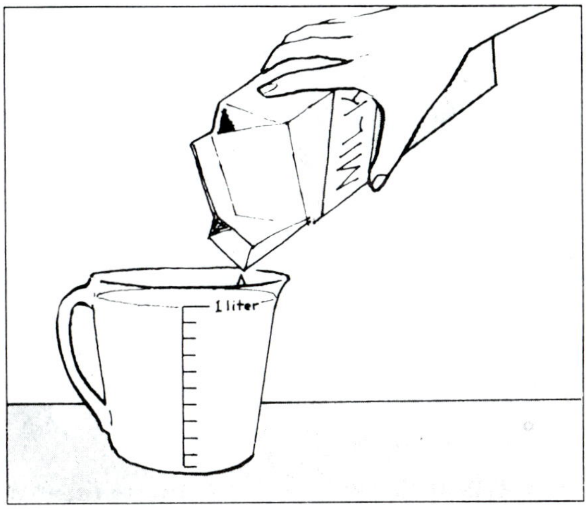

Figure B

Figure C

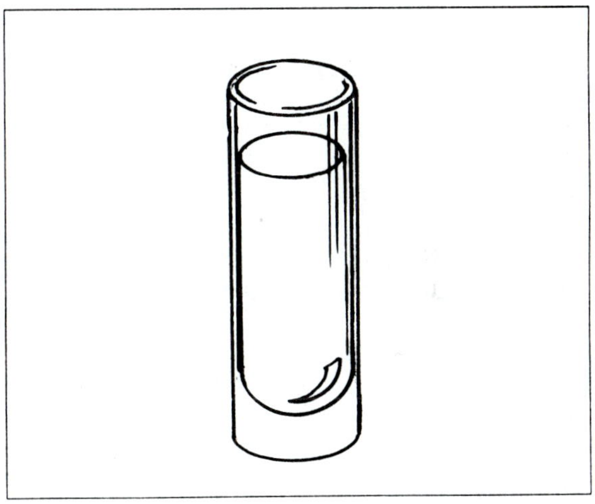

Figure D

Which figure shows that a liquid . . .

1. . . . takes up a definite amount of space? A B C D
2. . . . has mass? A B C D
3. . . . takes on the shape of its container? A B C D
4. . . . does not fill a container larger than itself? A B C D
5. . . . has no definite shape? A B C D

MULTIPLE CHOICE

In the space provided, write the letter of the word that best completes each statement.

_________ **1.** A liquid is matter because it

- **a)** is hard.
- **b)** has its own shape.
- **c)** has mass and takes up space.
- **d)** living things could not survive without it.

_________ **2.** The most common liquid on Earth is

- **a)** mercury.
- **b)** water.
- **c)** chicken soup.
- **d)** oxygen.

_________ **3.** All liquids

- **a)** are clear.
- **b)** flow.
- **c)** have odors.
- **d)** are flat on top.

_________ **4.** A liquid has a definite shape only when it is

- **a)** water.
- **b)** in a container.
- **c)** being poured.
- **d)** heavier than water.

_________ **5.** "Definite" means

- **a)** definition.
- **b)** fine.
- **c)** exact.
- **d)** changing.

TRUE OR FALSE

In the space provided, write "true" if the sentence is true. Write "false" if the sentence is false.

_________ **1.** A liquid has mass.

_________ **2.** The top of a liquid is always flat.

_________ **3.** A liquid has a shape of its own.

_________ **4.** All liquids have the same mass.

_________ **5.** Liquids come in many colors.

_________ **6.** Mercury is a solid.

_________ **7.** A liquid takes up a definite amount of space.

_________ **8.** A liquid can be made to have any shape.

_________ **9.** Water is the only liquid.

_________ **10.** Liquids and solids share ALL of the same properties.

Figure E

Figure F

COMPLETE THE CHART

Answer the questions by putting a YES or NO in the space provided.

		Have Mass	Take Up Space	Take Up A Definite Amount Of Space	Have Definite Shapes	Spreads Out To Take Shape of Its Container	Small Amounts Fill Large Container
1.	solids						
2.	liquids						

Use the chart to answer these questions. Write your answer in the space to the right of each question. Write *S* for solid, *L* for liquid.

1. Which states of matter have mass? ______
2. Which states of matter take up space? ______
3. Which of the states of matter take up a definite amount of space? ______
4. Which state of matter has a definite shape? ______
5. Which state of matter has no definite shape? ______
6. Which state of matter takes on the shape of its container? ______
7. Which state of matter does not take on the shape of its container? ______
8. Which states of matter do not always fill a container? ______

SCIENCE EXTRA

Cryogenics

Imagine surgeons performing "bloodless" operations. Today surgeons are able to perform some operations with very little bleeding by using extreme cold to kill diseased tissue. During surgery, they insert instruments with super-cooled tips into the diseased tissue. The cold tip kills the tissue. Healthy tissue surrounding the diseased tissue is shielded from the tip. This type of surgery, called cryosurgery, greatly reduces bleeding and complications.

Cryosurgery is just one use of cryogenics [kry-uh-JEN-iks]. Cryogenics is the study of extremely low temperatures. You probably think that 0°C—the freezing point of water—is cold. Temperatures in cryogenics are much colder. They range from about –120°C to –273°C, or absolute zero. Absolute zero is the coldest temperature a substance can reach.

Other medical uses of cryogenics include freezing tissues such as skin, the corneas of the eye, and blood. The tissues are then stored for future use. For example, corneas are stored in eye banks until they are needed to replace the diseased corneas of a patient.

Cryogenics also has other practical uses. In the food industry, it is used to store and preserve food quickly and easily. Foods are sealed in containers. The containers are sprayed with liquid nitrogen. Nitrogen normally exists as a gas. It must be super cooled to become a liquid. The super-cooled liquid nitrogen absorbs the heat of the foods.

Fuel industries use cryogenics to store natural gas reserves. The natural gas is stored as a super-cooled liquid. Gases also are cooled to liquids for storage on airplanes and in rockets. For example, as a safety measure liquid oxygen is carried on some airplanes. It can be changed to gas if the crew needs oxygen for any reason.

Someday, cryogenics may also play an important role in energy conservation. When certain metals are super-cooled they become superconductors. Superconductors conduct electricity without any resistance. Power lines made of superconducting materials could carry current over long distances without any loss of power from electrical resistance. These power lines would save large amounts of energy.

What are gases?

7

gas: phase of matter with no definite volume or shape

LESSON 7 | What are gases?

Can you see air? Why not? You cannot see air because it is a mixture of gases. Most gases are invisible. Like all matter, gases have mass, and take up space.

Even though it takes up space, a **gas** does not take up a definite amount of space. A gas is always moving. It will spread out in all directions to fill its container. A gas will even spread upward. In this way, gases are different from liquids and solids.

A gas has no definite shape. It takes on the shape of its container.

Many gases have odors. Some gases even have color. Some gases are poisonous. Chlorine (KLOWR-een) gas has all three of these characteristics. It is green, has an odor, and is poisonous.

Living thing need several gases to stay alive. Two very important gases for living things are oxygen [AHK-si-jun] and carbon dioxide [KAHR-bun dy-AHK-syd]. All living things need oxygen. Plants need carbon dioxide as well as oxygen. Without these two gases there would be no life.

WHAT DO THE PICTURES SHOW?

Each of these figures shows at least one property of gases. Look at each figure. Then answer the questions. Circle the letter of the figure that best answers each question.

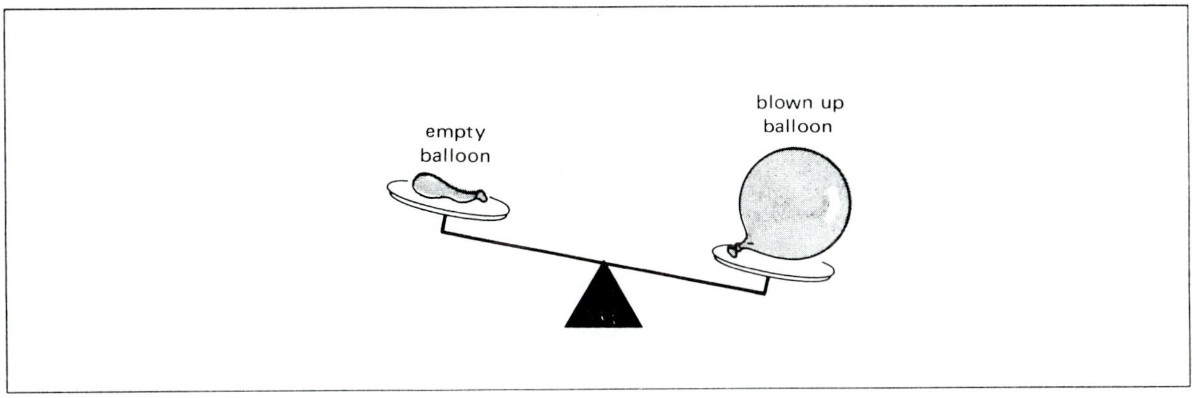

Figure A

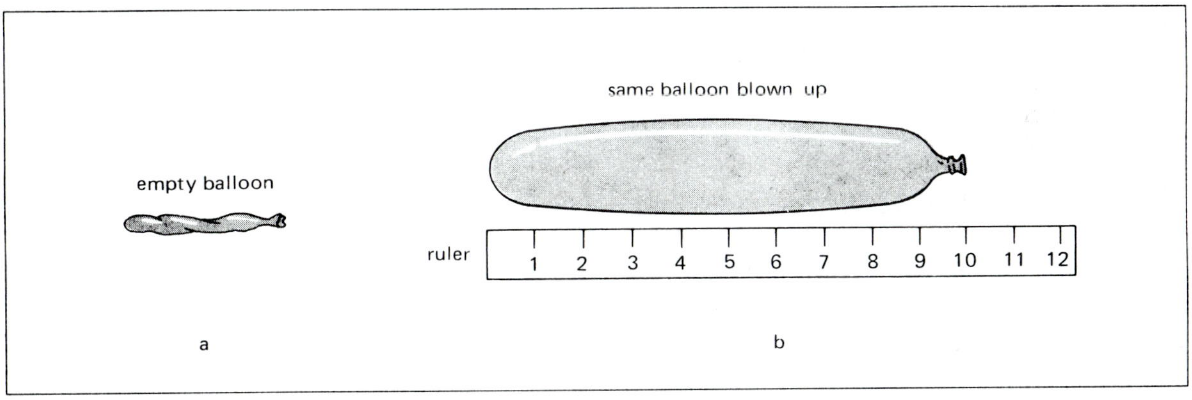

Figure B

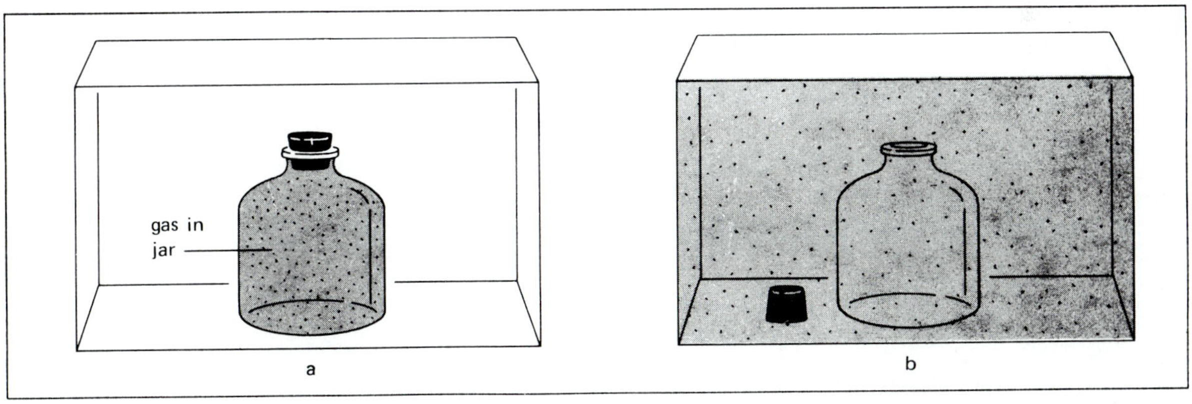

Figure C

Which picture shows that a gas . . .

1. . . . has mass? A B C

2. . . . takes up space? A B C

3. . . spreads out to fill its container? A B C

FILL IN THE BLANK

Complete each statement using a term or terms from the list below. Write your answers in the spaces provided. Some answers may be used more than once.

solids liquids gases plasmas matter

1. __________ have a definite shape.
2. __________ and __________ take up a definite amount of space.
3. __________ and __________ take the shape of their container.
4. __________ are very rare on Earth.
5. Some __________ are invisible.
6. __________, __________, __________, and __________ are the four states of __________.
7. Water and mercury are __________.
8. Oxygen and nitrogen are __________.
9. Rocks and metals are __________.
10. __________ will spread out to fill a large container.

TRUE OR FALSE

In the space provided, write "true" if the sentence is true. Write "false" if the sentence is false.

__________ 1. Oxygen is a gas.

__________ 2. All gases are invisible.

__________ 3. All gases have mass.

__________ 4. A gas has a definite shape.

__________ 5. A gas takes up a definite amount of space.

__________ 6. A gas will fill only part of a container.

__________ 7. Gases spread out only from side to side.

__________ 8. Every gas has an odor.

__________ 9. Animals need oxygen.

__________ 10. Plants need oxygen.

SHOWING THAT AIR TAKES UP SPACE

What You Need (Materials)

drinking glass
large bowl (or sink)
piece of paper
water

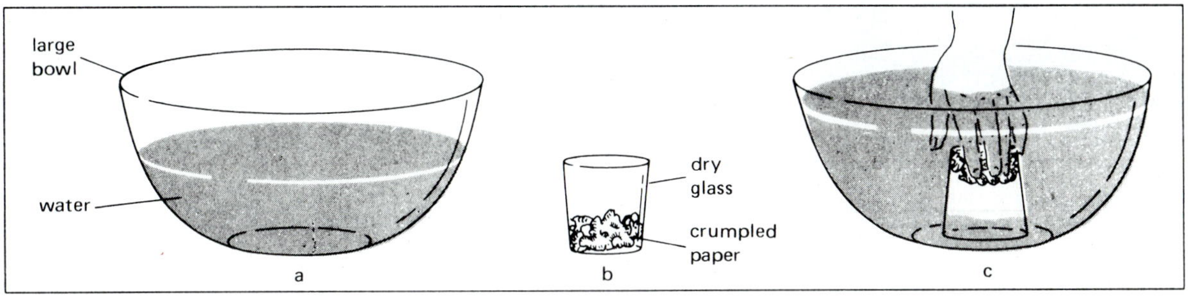

Figure D

How To Do the Experiment (Procedure)

1. Fill a large bowl (or your sink) halfway with water (Figure a).
2. Stuff a piece of paper into a small glass. Push it all the way to the bottom (Figure b).
3. Turn the glass upside down. Hold it straight. Put it into the bowl (Figure c). Hold it there for a short time. Then lift the glass out.
4. Look at the paper in the glass. Then take the paper out and feel it.

What You Learned (Observations)

1. Did the paper get wet? ______________________
2. Did the water get into the entire glass? ______________________
3. What stopped the water from filling the glass? ______________________
4. Can two things take up the same space at the same time? ______________________
5. This experiment shows that ______________________.
   air has weight, air takes up space

Something To Think About (Conclusions)

What do you think would happen if the bottom of the glass had a hole in it? ______________________

____

____

How would you explain that? ______________________

____

Figure E

Figure F

Figure G

COMPLETE THE CHART

Answer the questions by putting a YES or NO in the space provided.

	State	**A** Have Mass	**B** Take Up Space	**C** Take Up A Definite Amount Of Space	**D** Have Definite Shapes	**E** Spread Out To Take Shape Of Its Container	**F** Small Amounts Fill Large Containers	**G** Can Be Invisible
1.	solids							
2.	liquids							
3.	gases							

Answer the following questions. Decide which property or properties are correct for each question. Then write the letter(s) of the correct properties from the chart.

1. Which properties do all solids, liquids and gases have? ______

2. Which properties do only gases have? ______

3. Which property do solids and liquids have that gases do not have? ______

4. Which property do only solids have? ______

5. Which property do liquids and gases have that solids do not have? ______

REACHING OUT

Water is a liquid. When it boils it turns into steam. When it freezes it turns into ice. Using the properties in the chart above, write down what ice and steam are and why.

______

______

How does matter change from one state to another?

8

evaporate [i-VAP-uh-rayt]: change from a liquid into a gas
water vapor: water in the gas state

LESSON 8 How does matter change from one state to another?

Most kinds of matter can change from one state to another. Think about water. You know that liquid water can freeze and turn into solid water–ice. You have seen puddles of water seem to disappear. This happens when the liquid water turns into a gas–**water vapor**. The water **evaporates** [i-VAP-uh-rayts]. Clouds are made of water vapor. When the water vapor in the clouds cools, the water vapor turns back into a liquid and it rains.

Most matter can exist as a solid or a liquid or a gas. HEAT causes matter to change from one state to another.

RAISING THE TEMPERATURE: A rise in temperature changes some solids to liquids. Raising the temperature still higher will turn the liquid into a gas.

LOWERING THE TEMPERATURE: If you lower a gas's temperature, it may change into a liquid. Lowering the temperature more may change the liquid into a solid.

Different kinds of matter change their states at different temperatures. For example, ice changes to water at 0°C (32°F). But you have to get iron much hotter than that to turn it to a liquid.

The fact that matter can change from one state to another is important to us. If we could not melt metals to liquid and turn them back into solids, we would not be able to make many products, like automobiles and frying pans.

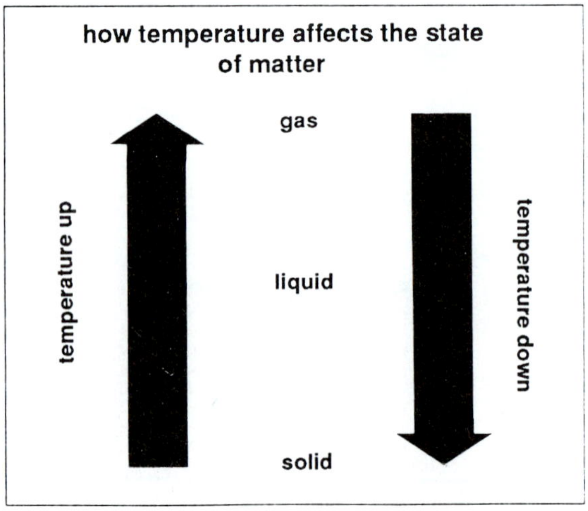

BOILING AND FREEZING TEMPERATURES OF WATER

Freezing Temperature	0 degrees Celsius 32 degrees Fahrenheit	(0°C) (32°F)
Boiling Temperature	100 degrees Celsius 212 degrees Fahrenheit	(100°C) (212°F)

Figure A

Figure B

Using Figures A and B answer the following questions.

1. Is water found only in one state? ________________

2. In what states is water found? ____

3. What in the solid state is called ________________ .

4. Water in the gas state is called ________________ .

5. For liquid water to change to ice, its temperature must ________________ .
   rise, drop

6. For ice to melt, its temperature must ________________ .
   rise, drop

7. For water to change to a gas, its temperature must ________________ .
   rise, drop

8. Water that changes to a gas is said to ________________ .
   evaporate, melt

9. For water vapor to change back to a liquid, its temperature must ________________ .
   rise, drop

10. Water changes into ice at ________________ degrees Celsius.
    0, 100

TWO COMMON GASES

Study each of the pictures below. Then answer the questions.

Figure C

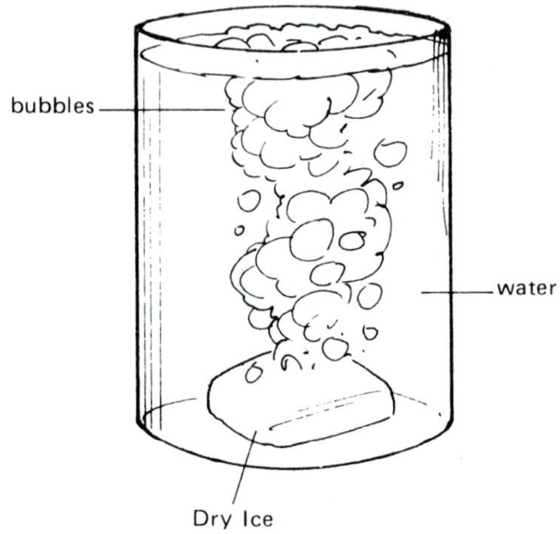

Figure D

Carbon dioxide usually is a gas. However, it can be made into a solid. Some people call this solid "dry ice." It is called this because it does not melt! It changes to a gas. It evaporates! Dry ice is used to keep things cold. It is usually colder than regular ice.

1. Why is dry ice used in ice cream carts? ______________________________

____

____

2. What must be done to carbon dioxide gas to change it to a solid? ______________

____

3. Which is colder, dry ice or carbon dioxide gas? ______________________

If dry ice is placed in water we see bubbles rise.

4. These bubbles are made of ______________ in the ______________ state.

5. What happens to the size of the dry ice as the bubbling goes on? ______________

____

Figure E

The Space Shuttle uses liquid hydrogen and liquid oxygen for fuel. When the two liquids are burned together, they form water. The liquid oxygen and liquid hydrogen are stored separately in the huge orange external tank.

1. What is the usual state of oxygen and hydrogen? ______________________

2. What must be done to oxygen and hydrogen to change them to liquids? ______________________

   ____

3. Before lift-off, is the external tank cold or hot? ______________________

4. As the water produced in the engine is heated, what will happen to the water's temperature? ______________________

5. What has happened to the water by the time it leaves the engine? ______________________

   ____

MATCHING

Match each term in Column A with its description in Column B. Write the correct letter in the space provided.

	Column A	Column B
______	1. 100°C (212°F)	a) water in the gaseous state
______	2. dry ice	b) water in the solid state
______	3. water vapor	c) freezing temperature of water
______	4. ice	d) carbon dioxide in the solid state
______	5. 0°C (32°F)	e) boiling temperature of water

FILL IN THE BLANK

Complete each statement using a term or terms from the list below. Write your answers in the spaces provided. Some answers may be used more than once.

gas	freezes	solid
100°C (212°F)	evaporates	water vapor
dry ice	liquid	0°C (32°F)

1. At room temperature, water is a ________________ .
2. If liquid water becomes cold enough, it changes to ice. Ice is a ________________ .
3. When water changes to ice, we say that it ________________ .
4. The temperature at which water freezes is ________________ .
5. If the temperature of water rises, it changes to a ________________ .
6. Water in the gas state is called ________________ .
7. When water changes to a gas, we say that it ________________ .
8. ________________ changes directly from a solid to a gas.
9. When a gas cools, it may change to a ________________ .
10. The temperature at which water boils is ________________ .

COMPLETE THE CHART

Answer the questions by putting a YES or NO in the space provided.

	Change Of State	Rise In Temperature?	Drop In Temperature?
1.	solid to liquid		
2.	liquid to gas		
3.	gas to liquid		
4.	liquid to solid		
5.	solid to gas		
6.	gas to solid		

What are atoms?

9

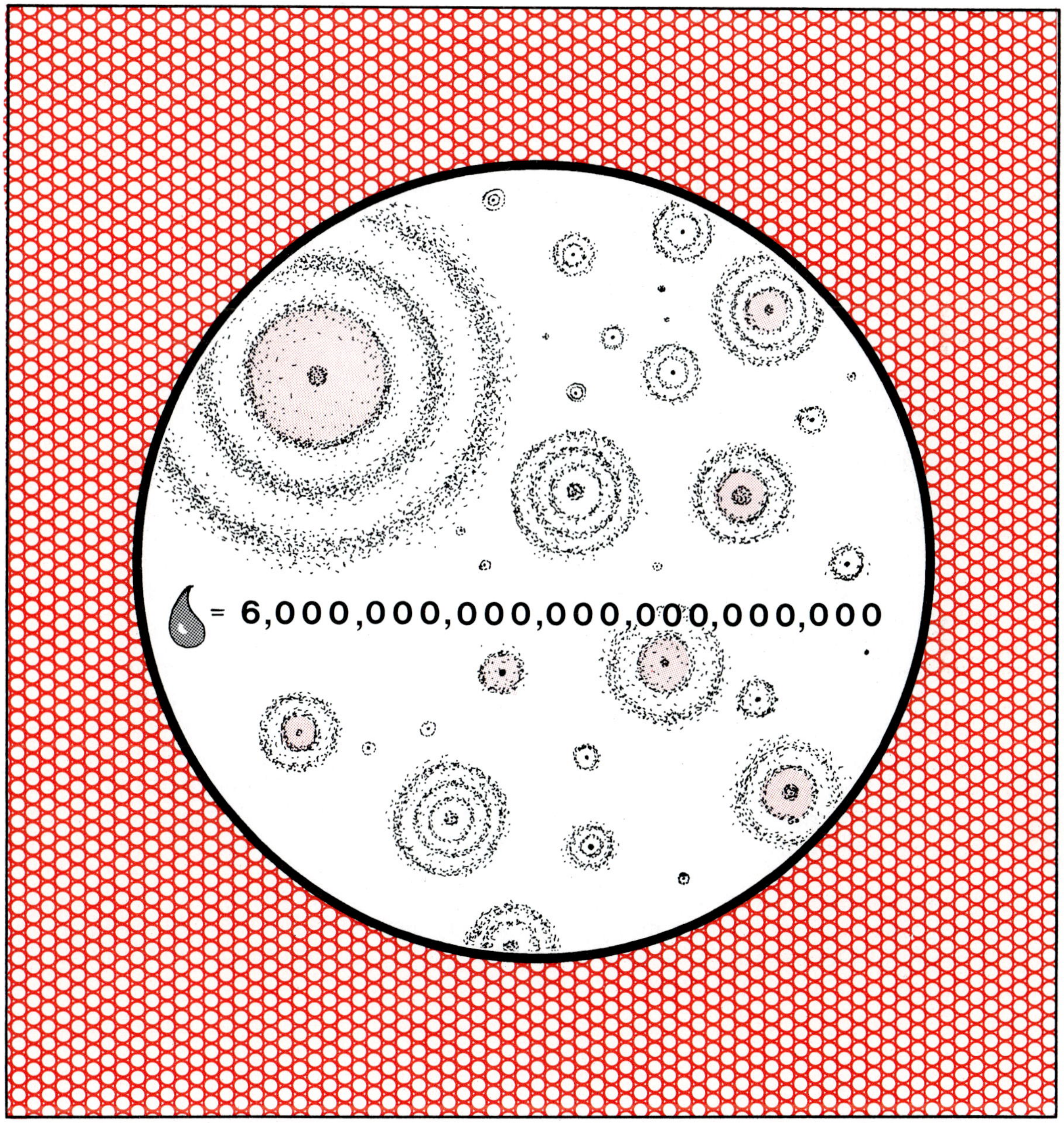

atom: smallest particle of a type of matter that has all of the same characteristics as that type of matter

LESSON 9 | What are atoms?

What is the smallest thing you can think of? A single grain of sand? A particle of dust?

Now try to imagine something so small that you would need millions of them to make one grain of sand! Imagine something so small that you cannot see it—not even with the most powerful microscope.

There is something that small: the **atom**. All matter is made up of atoms. All solids, liquids, gases, and plasmas are made up of these tiny particles.

Atoms are matter. One atom take up space—very, very, very little space. An atom also has mass—very, very, very little mass.

How small is the atom? Atoms are so small that in just one drop of water, there are about six sextillion atoms.

That's 6,000,000,000,000,000,000,000 atoms!!!

If you tried to count to six sextillion it would take you about one hundred trillion years—If you counted fast!

The idea of the atom is far from new. Many years ago, before there were any "real" scientists, there were philosophers [fi-LAHS-uh-furz]—people who did mental "investigations." They worked with ideas. Over 2,000 years ago, a Greek philosopher named Democritus [di-MAHK-ruh-tus] had the idea that all matter was made up of tiny parts. He believed that these parts could not be divided or destroyed. He named them *atoms*. In Greek, *atomos* means "indivisible"[in-di-VIS-uh-bul].

Democritus could not prove his idea. He couldn't even test it. So, it remained just an idea for many years. Today, scientists have proven that many of Democritus's ideas were correct. Everyday, more and more is discovered about the atom.

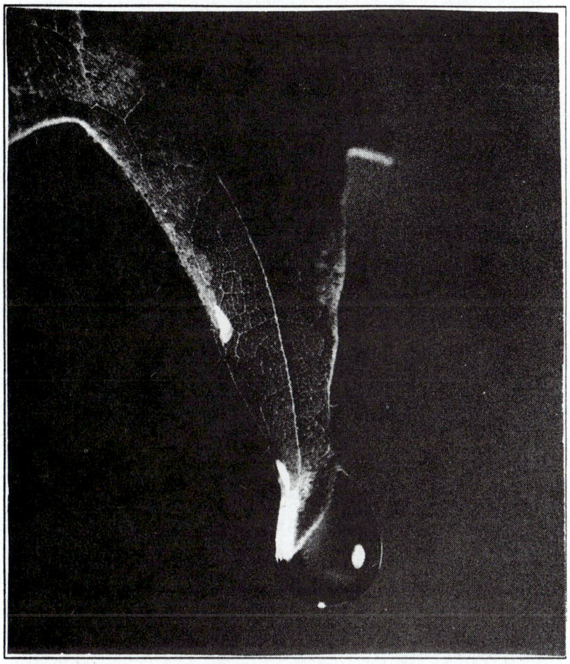

Figure A

Suppose each of those atoms were a drop of water. How much water would that be? It would be six sextillion drops of water.

Figure B

That's more water than all the water that passes over Niagara Falls in 2000 years.

Figure C

That's enough water to fill about six billion Empire State Buildings.

Figure D

That's almost twice as much water as there is in the Mediterranean Sea.

TRUE OR FALSE

In the space provided, write "true" if the sentence is true. Write "false" if the sentence is false.

_________ **1.** An atom is very large.

_________ **2.** Democritus named the atom.

_________ **3.** Solids are made of atoms.

_________ **4.** Liquids are made of atoms.

_________ **5.** Gases are not made of atoms.

MATCHING

Match each term in Column A with its description in Column B. Write the correct letter in the space provided.

	Column A	Column B
_________	**1.** Democritus	**a)** a very small particle
_________	**2.** matter	**b)** Greek philosopher who named the atom
_________	**3.** philosophers	**c)** made up of atoms
_________	**4.** *atomos*	**d)** people who think about things
_________	**5.** atom	**e)** Greek word for "indivisible"

WHICH IS SMALLEST? WHICH IS BIGGEST?

Each group of words or terms below can be arranged by size. Write them in the correct order in the spaces below each group.

1. a piece of dust   an atom   an elephant

_________________ _________________ _________________

smallest   largest

2. the tip of a pin   a dime   an atom

_________________ _________________ _________________

smallest   largest

3. an atom   a rock   a pebble

_________________ _________________ _________________

smallest   largest

MODERN ATOMIC THEORY

Figure E *John Dalton*

Figure F *Synchrotron*

In the early 1800s, an English chemist named John Dalton described his ideas about matter. Dalton's ideas were based on many scientific experiments and observations. The ideas formed a theory that led to our modern atomic theory.

You may wonder how we could know anything about a particle of matter that is too small to see and almost too small to measure. Scientists have learned how to study atoms. They study atoms by studying how matter behaves. They use very complicated equipment. However, you can learn about atoms by studying what scientists have learned.

The present atomic theory states:

1. All elements are made up of tiny particles called atoms.
2. Atoms of a given element are alike.
3. Atoms of different elements are different.
4. Chemical changes take place when atoms link up with, or separate from, one another.
5. Atoms are not created or destroyed by chemical change.

Democritus was on the right track over 2000 years ago. However, one important part of his idea has been proven wrong. Atoms *are* divisible. In fact, the "splitting" of the atom is the basis for nuclear [NEW-klee-ur] or atomic, energy.

FILL IN THE BLANK

Complete each statement using a term or terms from the list below. Write your answers in the spaces provided. Some answers may be used more than once.

alike	John Dalton	atoms
created	six sextillion	indivisible
different	small	destroyed
Democritus	2000	

1. The atom was first thought of by a man named ________________ more than ________________ year ago.
2. In Greek, the word *atomos* means ________________ .
3. Matter that is indivisible cannot be ________________ .
4. An English chemist named ________________ presented a modern atomic theory.
5. All elements are made of ________________ .
6. Atoms of a given element are all ________________ .
7. Atoms can not be ________________ or ________________ by chemical change.
8. Atoms of ________________ elements are different.
9. Atoms are so ________________ that we can not see them.
10. There are ________________ atoms in a drop of water.

REACHING OUT

Why did it take 2000 years for scientist to confirm some of Democritus's ideas about atoms.

____

____

____

____

____

____

What are the parts of an atom?

10

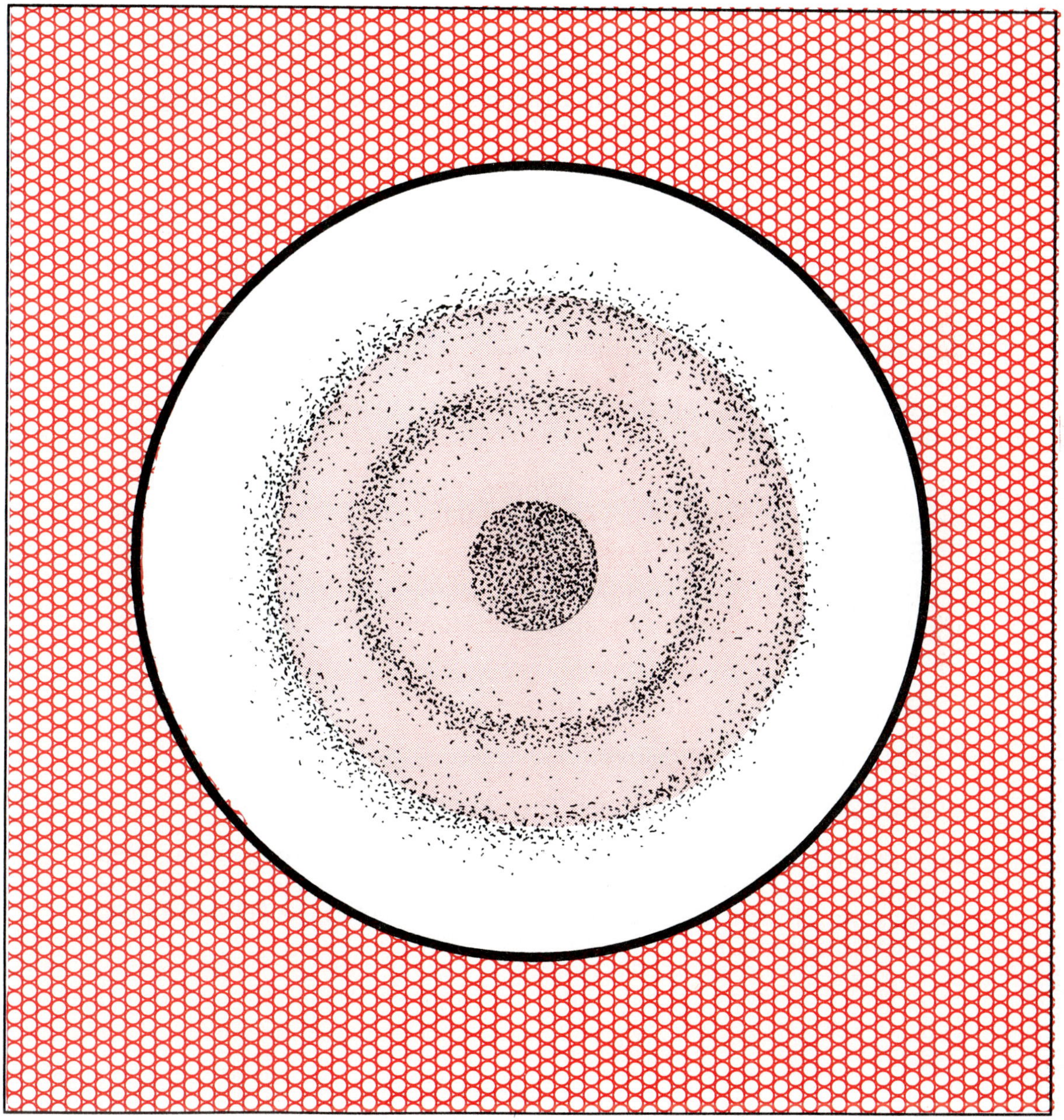

electron [i-LEK-trahn]: a part of the atom that has a negative electrical charge; orbits the nucleus

neutron [NEW-trahn]: a part of the atom that has neither a positve or negative charge; is found in the nucleus

nucleus [NEW-klee-us]: central part of an atom, which contains neutrons and protons

proton [PROH-tahn]: a part of the atom that has a positive charge; is found in the nucleus

LESSON 10 What are the parts of an atom?

People once thought that the atom was the smallest particle of matter in the universe. However, scientists now know that atoms are made up of even smaller parts. There are three different kinds of particles. They are: **protons** [PRO-tahnz], **neutrons** [NEW-trahnz], and **electrons** [i-LEK-trahnz].

Most of the mass of an atom is found in the central part of the atom, called the **nucleus** [new-KLEE-us]. The nucleus of an atom is made up of protons and neutrons. These particles are packed very tightly together in the nucleus.

Electrons are found outside the nucleus. They circle the nucleus very, very quickly. Electrons are very small and have almost no mass. The number of electrons in an atom is always equal to the number of protons in the nucleus of that atom.

Scientists have discovered that protons, electrons, and neutrons have different **charges**. You probably know that the word "charge" has something to do with electricity.

There are two kinds of charges. There are positive (plus) charges and negative (minus) charges. By studying atoms, scientists have learned that:

- PROTONS have positive (+) charges.
- ELECTRONS have negative (–) charges.
- NEUTRONS have no charges. They are neutral.

Since atoms have the same number of protons and electrons, the number of positive charges equals the number of negative charges. The opposite charges cancel each other out. Therefore, THE WHOLE ATOM HAS NO OVERALL CHARGE.

ATOMIC DIAGRAMS

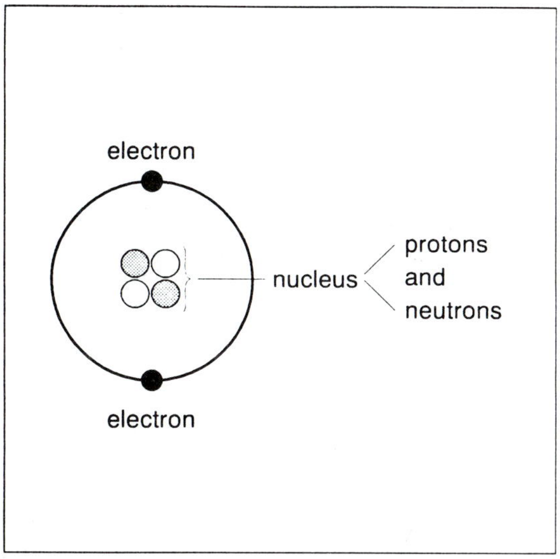

Figure A

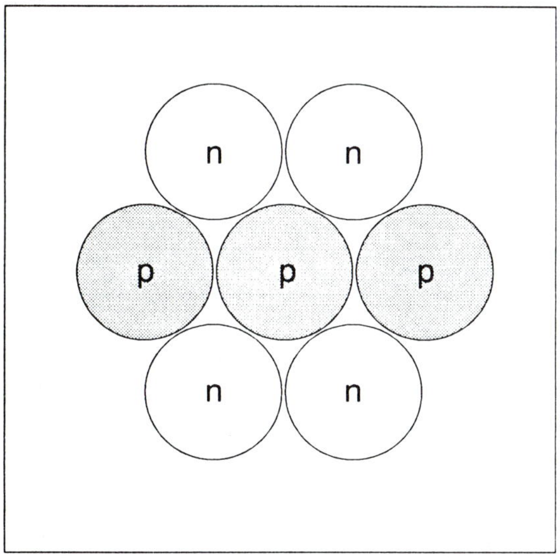

Figure B

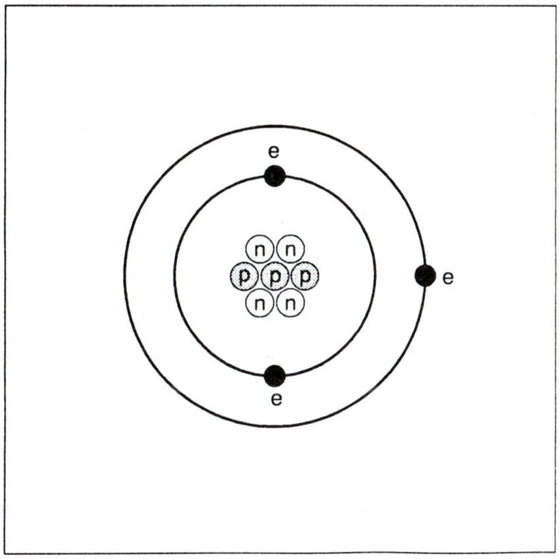

Figure C

The table below tells where the parts of the atom are found and what the charge of each part is.

Name of part	Where it is found	Charge
proton	inside the nucleus	+
neutron	inside the nucleus	0
electron	outside the nucleus	–

Figure B shows the center of a <u>lithium</u> atom. The center of an atom is called it nucleus.

1. Name the parts that make up a nucleus.

____

2. In the diagram, each "p" stands for

______________ ; each "n" stands for

a ______________ .

3. How many protons are in a lithium nucleus?

____

4. How many neutrons are in a lithium nucleus? ______________

Figure C shows a <u>full</u> lithium atom.

5. Label the nucleus on the diagram.

6. Draw in a small "e" next to each electron.

7. How many electrons does a lithium atom have? ______________

8. How many positive charges are in the atom?

____

9. How many negative charges are in the atom?

____

10. What is the overall charge of the atom? ____

INTERPRETING ATOMIC DIAGRAMS

Below and on the following page are diagrams of six different atoms. In the spaces provided to the right of each diagram, fill in the number of protons, neutrons, electrons, positive charges, negative charges, and the overall charge of each atom.

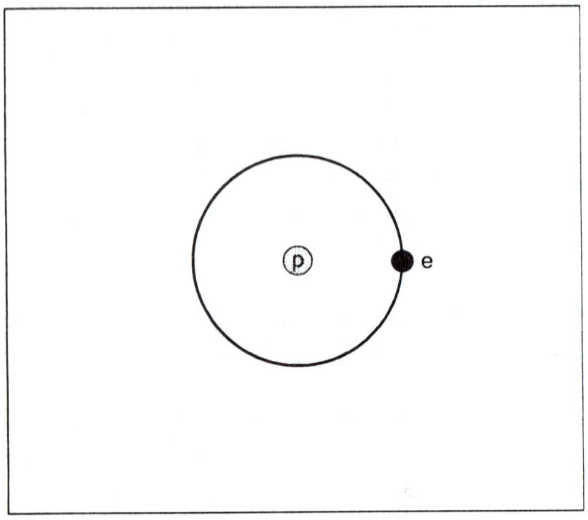

Protons ______________

Neutrons ______________

Electrons ______________

Positive charge ______________

Negative charge ______________

Overall charge ______________

Figure D *Hydrogen*

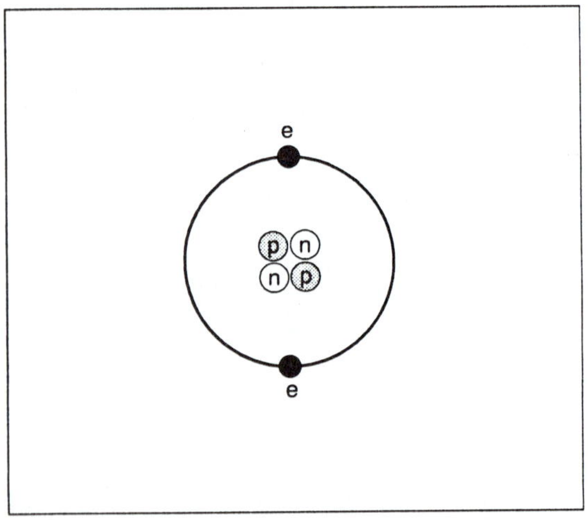

Protons ______________

Neutrons ______________

Electrons ______________

Positive charge ______________

Negative charge ______________

Overall charge ______________

Figure E *Helium*

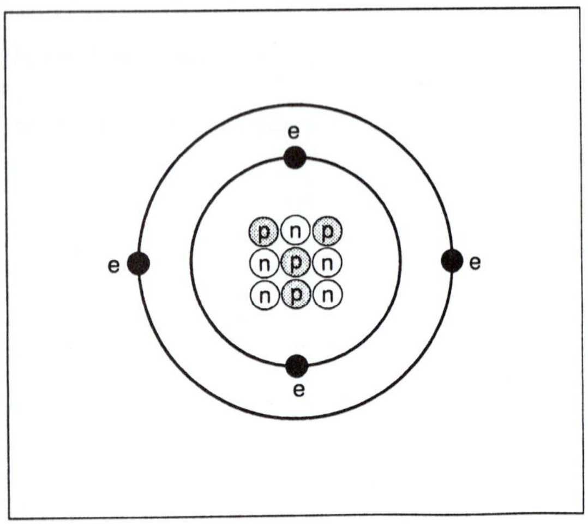

Protons ______________

Neutrons ______________

Electrons ______________

Positive charge ______________

Negative charge ______________

Overall charge ______________

Figure F *Beryllium*

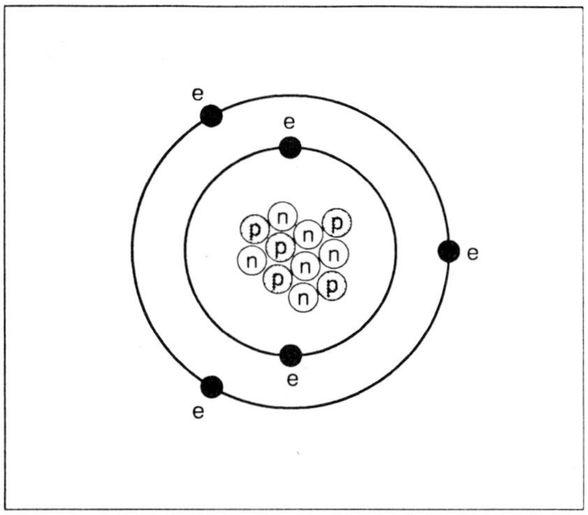

Protons ________________

Neutrons ________________

Electrons ________________

Positive charge ________________

Negative charge ________________

Overall charge ________________

Figure G *Boron*

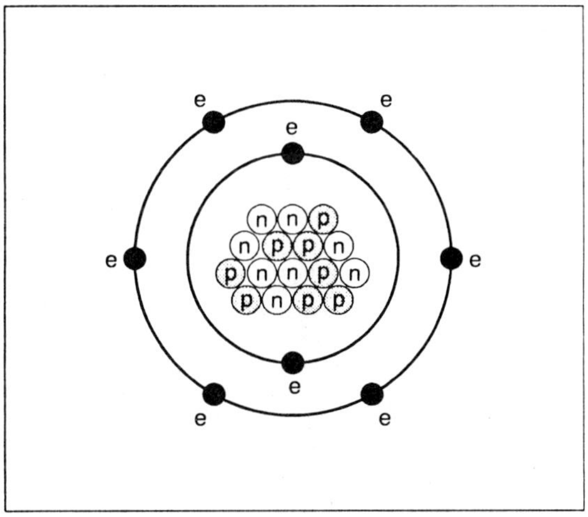

Protons ________________

Neutrons ________________

Electrons ________________

Positive charge ________________

Negative charge ________________

Overall charge ________________

Figure H *Oxygen*

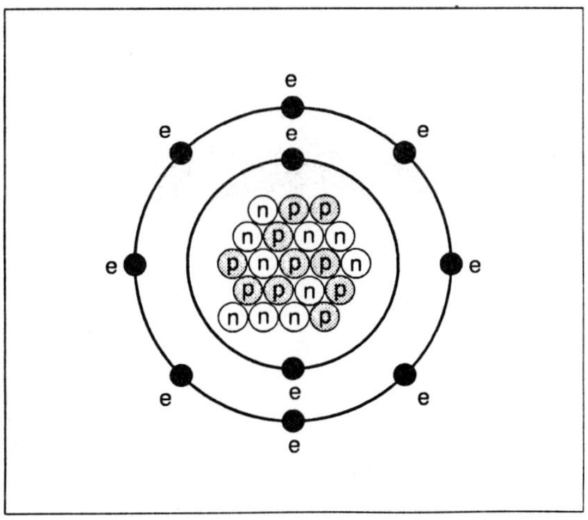

Protons ________________

Neutrons ________________

Electrons ________________

Positive charge ________________

Negative charge ________________

Overall charge ________________

Figure I *Neon*

FILL IN THE BLANK

Complete each statement using a term or terms from the list below. Write your answers in the spaces provided. Some answers may be used more than once.

outside	neutrons	same	cancel out
protons	atoms	negative	electrons
nucleus	smaller	no	positive

1. All matter is made of tiny parts called ________________ .
2. The center part of an atom is called the ________________ .
3. A nucleus is made up of ________________ and ________________ .
4. Electrons are found ________________ the nucleus.
5. Electrons are ________________ than protons or neutrons.
6. The main parts of an atom are ________________ , ________________ , and ________________ .
7. Since protons have a ________________ charge, and neutrons have ________________ charge, the nucleus will have a ________________ charge.
8. Electrons have a ________________ charge.
9. An atom has the ________________ number of protons and electrons.
10. The plus and minus charges of an atom ________________ each other.

TRUE OR FALSE

In the space provided, write "true" if the sentence is true. Write "false" if the sentence is false.

________ 1. A proton is found outside the nucleus.

________ 2. A proton has a negative charge.

________ 3. A neutron has a positive charge.

________ 4. An electron has a negative charge.

________ 5. An electron is found inside the nucleus.

How are electrons arranged in an atom?

11

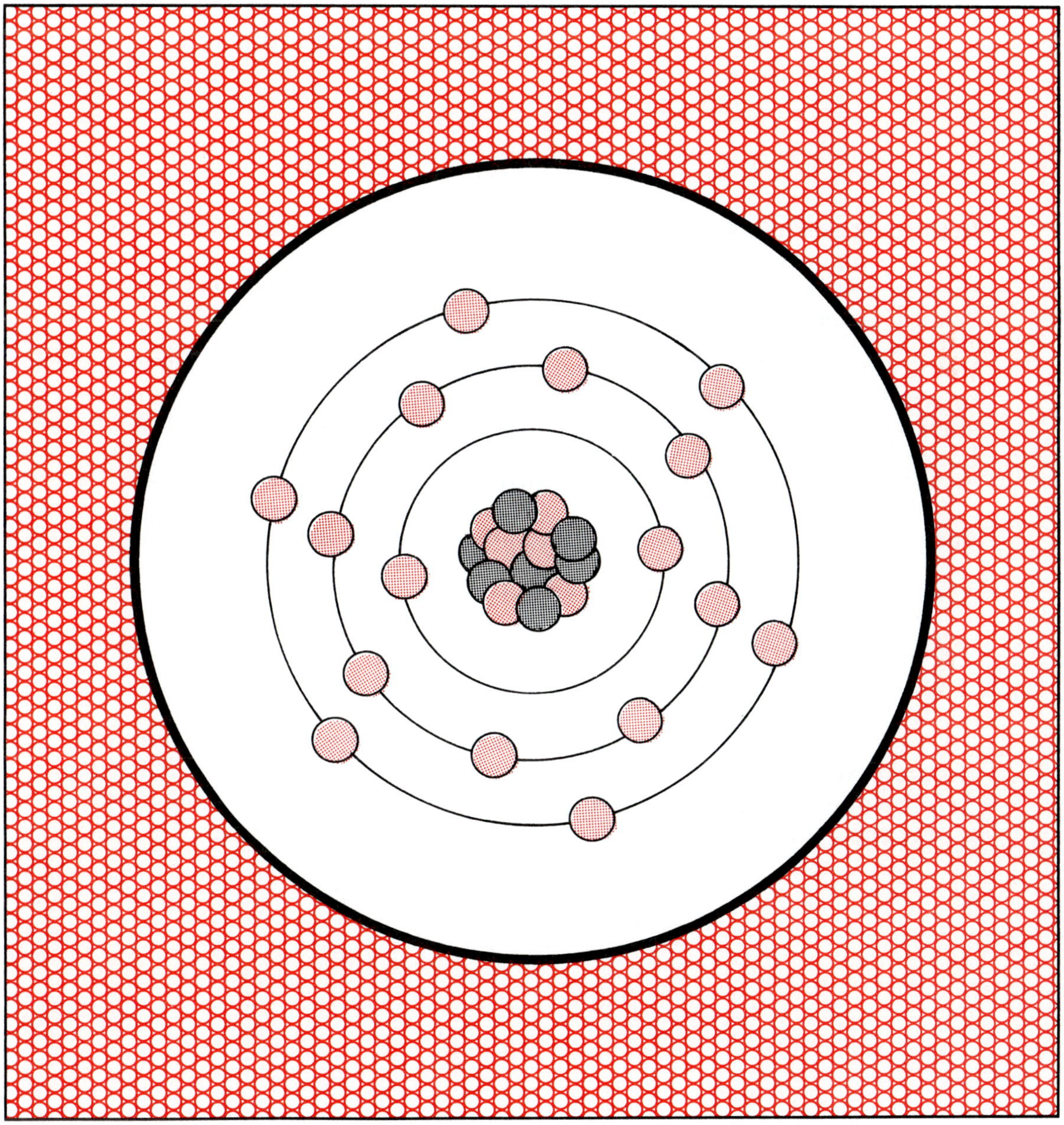

shells: energy levels in which electrons are arranged around the nucleus

LESSON 11 How are electrons arranged in an atom?

In the past, scientists believed that electrons circled the nucleus the same way the planets circled the sun. Today, however, scientists know that there is no exact path of an electron. The quick moving electrons form a "cloud" around the nucleus.

In the modern atomic theory, electrons are arranged into energy levels, or **shells**. Each electron shell is labelled with a capital letter. The first shell is the "K" shell. It is the shell closest the the nucleus. The "K" shell has the least amount of energy. The next shell is the "L" shell. After the "L" shell, comes the "M" shell, and so on.

Each shell can only hold a certain number of electrons.

- The "K" shell can hold up to 2 electrons.
- The "L" shell can hold up to 8 electrons.
- The "M" shell can hold up to 18 electrons.

The number of shells an atom has depends upon the number of electrons the atom has. In general, each shell must have its full number of electrons before a new shell starts. If there are more electrons than a shell can hold, a new shell starts.

MORE ABOUT ELECTRONS

Each picture below shows how electrons are arranged in certain atoms. Look at each diagram. Then answer the questions.

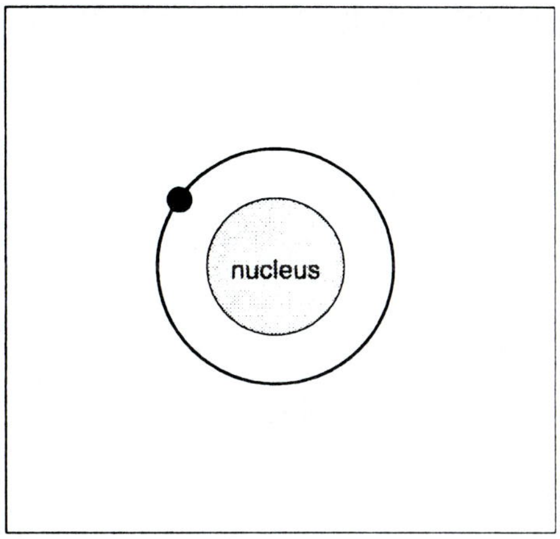

Figure A *Hydrogen*

A. Hydrogen is the simplest atom. It has only one electron.

1. How many electron shells does a hydrogen atom have? ______________________
2. What is this shell called? ______________
3. Is this shell complete? ________________

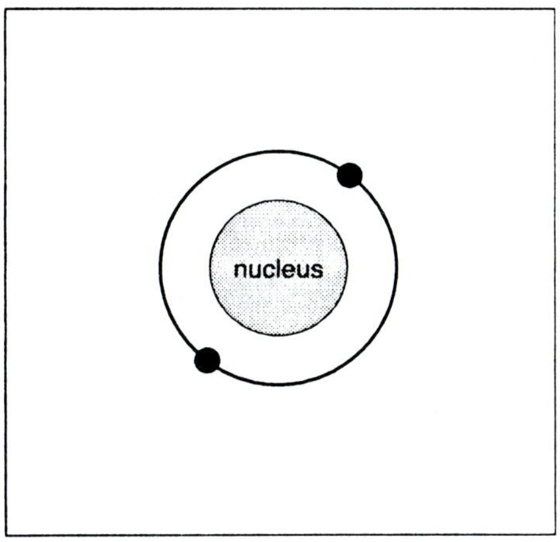

Figure B *Helium*

B. A helium atom has two electrons.

1. How many electron shells does a helium atom have? ______________________
2. What is this shell called? ______________
3. Is the shell full? ________________

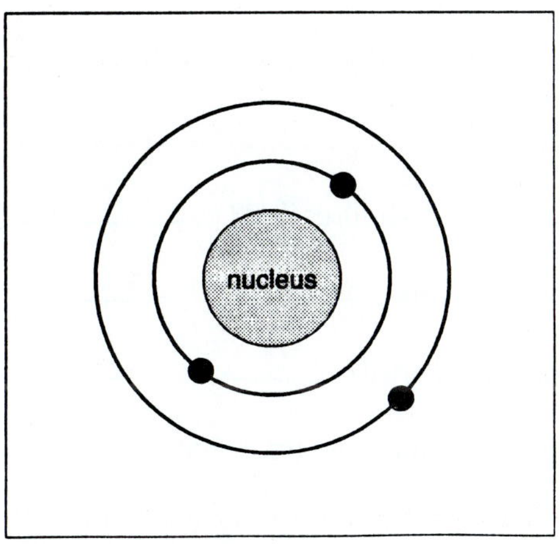

Figure C *Lithium*

C. A lithium atom has three electrons.

1. How many electron shells does a lithium atom have? ______________________
2. What are these shells called? ____________
3. Which one is the inner shell? ____________
4. Is the inner shell full? ______________
5. Which is the outer shell? ______________
6. Is the outer shell full? ______________

WHAT DO THE PICTURES SHOW?

The pictures below show how electrons are arranged in two atoms. Look closely at each figure. Then answer the questions.

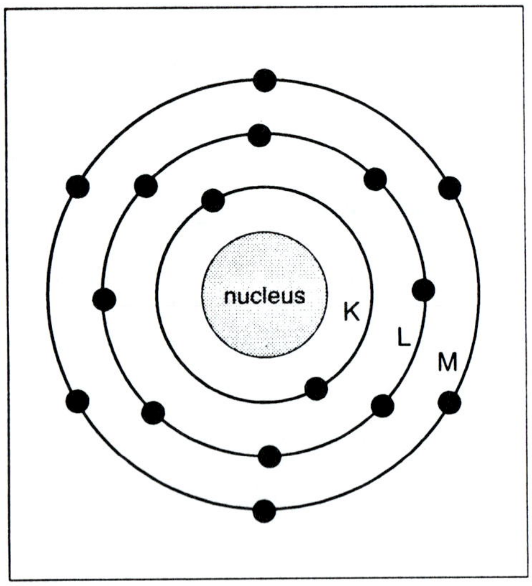

Figure D *Sulfur*

1. How many electron shells does sulfur have?

 ____

2. What is the first shell called? ______________
3. How many electrons does this shell have? ____
4. Is this shell full? ______________________
5. What is the second shell called? ____________
6. How many electrons does it have? __________
7. Is this second shell full? _________________
8. What is the third shell called? ____________
9. How many electrons does the third shell have?

 ____

10. Is this shell full? ______________________
11. How many electrons does sulfur have? ______

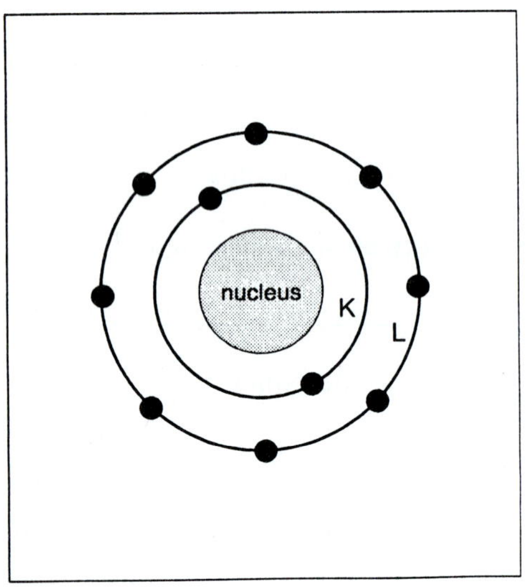

Figure E *Neon*

1. How many electron shells does neon have? ___
2. What is the first shell called? ______________
3. How many electrons does this shell have? ____
4. Is this shell full? ______________________
5. What is the second shell called? ____________
6. How many electrons does it have? __________
7. Is this second shell full? _________________
8. How many electrons does neon have? _______

MATCHING

Match each term in Column A with its description in Column B. Write the correct letter in the space provided.

	Column A	Column B
________	**1.** shell	**a)** closest shell
________	**2.** 5 electrons	**b)** way electrons circle nucleus
________	**3.** cloud	**c)** needs three shells
________	**4.** 12 electrons	**d)** energy level
________	**5.** "K" shell	**e)** needs two shells

HOW MANY SHELLS?

How many shells are needed for each of the following? Write your answer in the space.

1. How many shells do 2 electrons need? ________

2. 4 electrons need ________ shells.

3. 10 electrons need ________ shells.

4. 16 electrons need ________ shells.

5. 20 electrons need ________ shells.

FILL IN THE ELECTRONS

Draw in the electrons on their proper shells. Make a small ball [●] to show an electron. The first one is done for you.

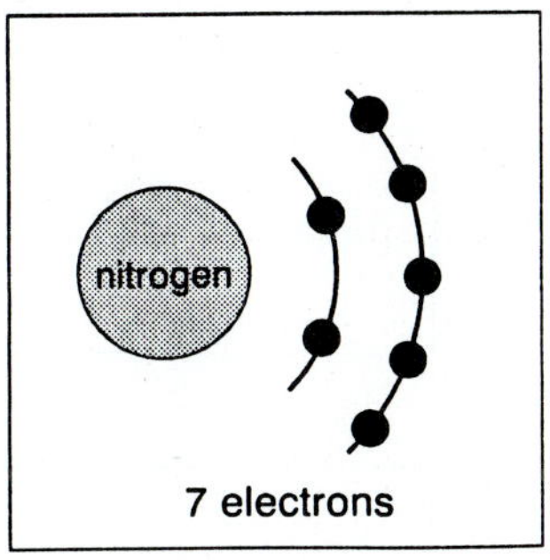

Figure F

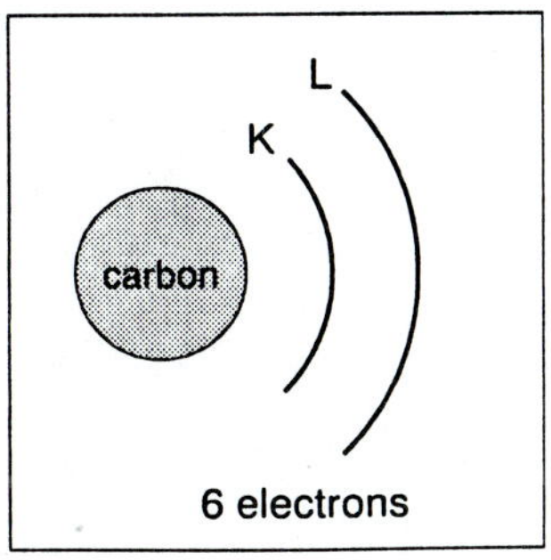

Figure G

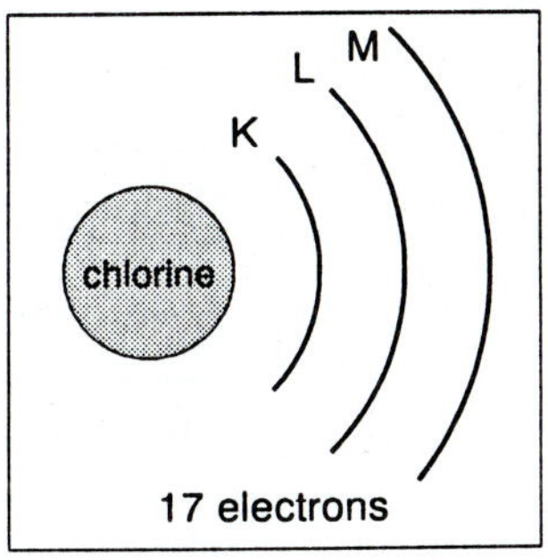

Figure H

TRUE OR FALSE

In the space provided, write "true" if the sentence is true. Write "false" if the sentence is false.

_________ **1.** Every atom has at least one electron.

_________ **2.** Every atom has at least two shells.

_________ **3.** The first electron shell is the "L" shell.

_________ **4.** The "L" shell can hold 8 electrons.

_________ **5.** The "L" shell always has 8 electrons.

_________ **6.** If there is an "L" shell, it means that the "K" shell is full.

_________ **7.** A full "K" shell has three electrons.

_________ **8.** A helium atom which has 2 electrons is a "full" atom.

_________ **9.** A shell must be full before a new shell is started.

_________ **10.** An atom with 10 electrons has 2 shells.

COMPLETE THE CHART

Complete the chart by filling in the missing information.

	Atom	Number of Electrons	Number of electrons in each shell			Is last shell complete? (Yes or No)
			K	L	M	
1.	Magnesium	12	2	8	2	No
2.	Carbon	6				
3.	Oxygen	8				
4.	Helium	2				
5.	Neon	10				
6.	Aluminum	13				
7.	Chlorine	17				
8.	Phosphorus	15				
9.	Argon	18				
10.	Beryllium	4				

What is the difference between atomic mass and atomic number?

12

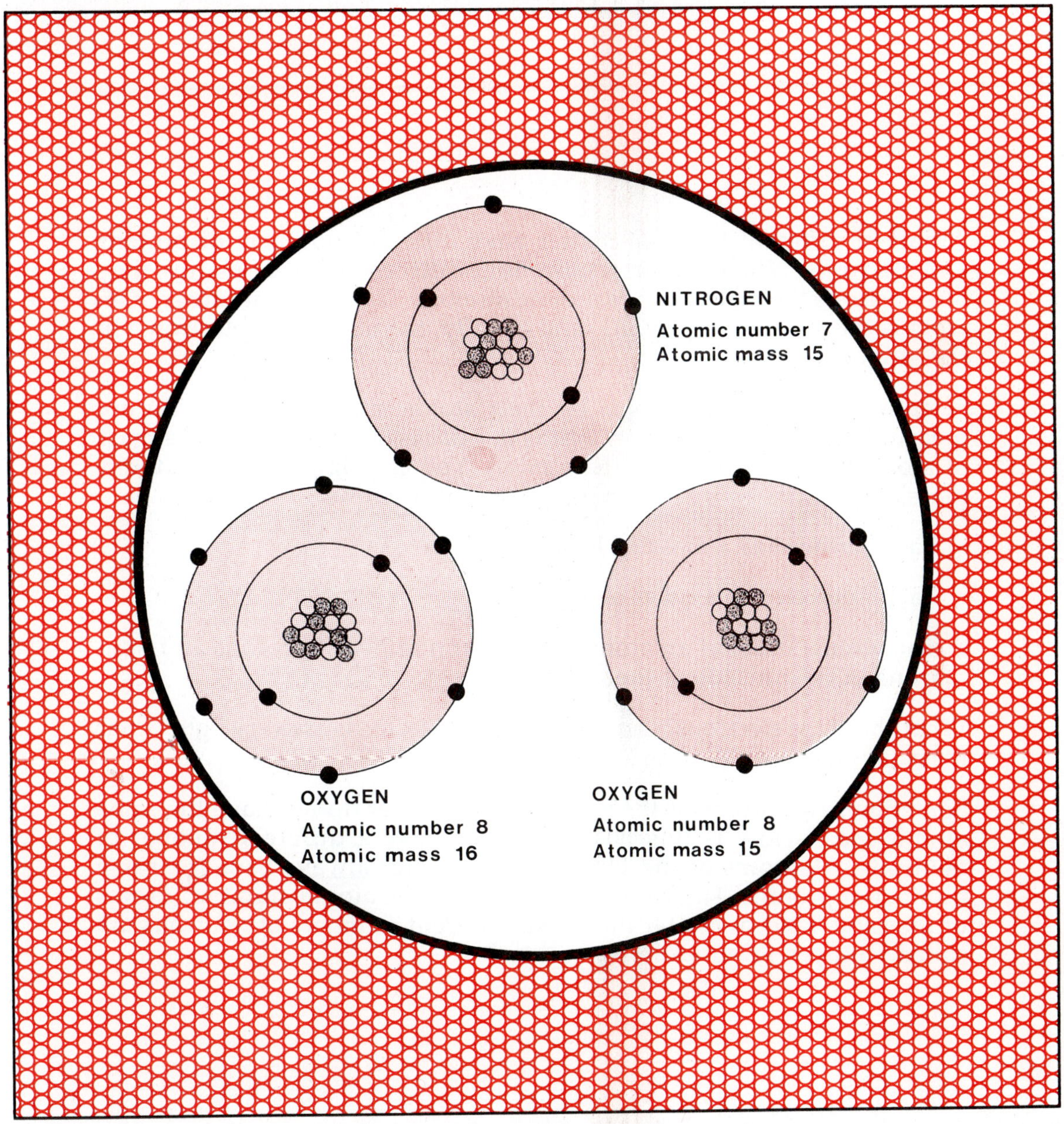

atomic mass: total number of protons and neutrons in the nucleus of an atom
atomic mass units: units used to measure the mass of atoms
atomic number: number of protons in the nucleus of an atom
isotopes [Y-suh-tohps]: two or more atoms with the same atomic number but different atomic masses

LESSON 12 | What is the difference between atomic mass and atomic number?

Atoms of different kinds of matter have different numbers of protons and electrons. When scientists talk about different kinds of matter, they often refer to the matter by it's atomic number. The **atomic number** of an atom is the number of protons (and usually, the number of electrons) in the atom.

Scientists also describe atoms by their atomic mass. Scientists do not measure the mass of atoms in grams or ounces. They measure the mass of atoms in **atomic mass units (a.m.u.)**. You can figure out the atomic mass of an atom by using the following information:

- Each proton has a mass of 1 a.m.u.
- Each neutron has a mass of 1 a.m.u.
- The **atomic mass** of an atom is the total number of protons and neutrons in the nucleus of the atom.

What about the electrons? Don't they count? Electrons are <u>very</u> light. Their mass is not counted in the atomic mass.

Sometimes two atoms of the same kind of matter do not have the same atomic mass. How is this possible? They have a different number of neutrons. All atoms of the same kind of matter <u>always</u> have the same number of protons. Thus, they all have the same atomic number. Atoms of the same kind of matter that have different numbers of neutrons are called **isotopes** [Y-suh-tohps].

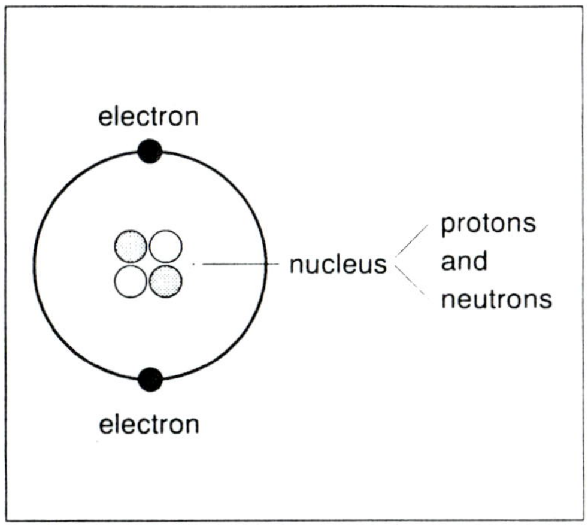

Figure A

NUCLEUS

PROTONS + NEUTRONS = ATOMIC MASS

Each proton has a value of <u>one</u>.

Each neutron has a value of <u>one</u>.

ATOMIC MASS, PLEASE

The diagrams below show six different atoms. Look at each one closely. Find the atomic mass of each atom. Write your answer in the space below the diagram.

Remember: atomic mass = protons + neutrons

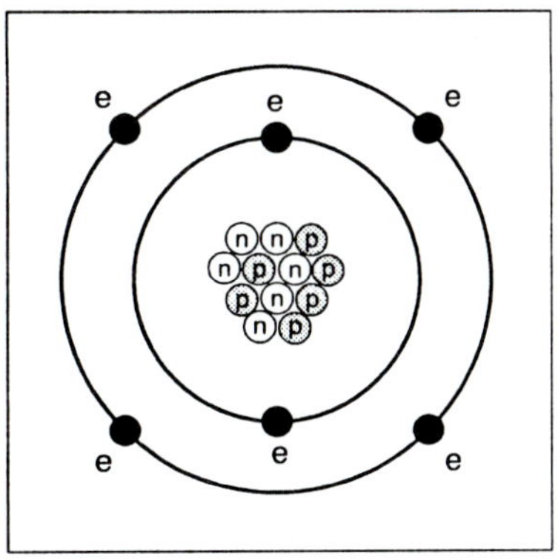

Figure B

Atomic Mass ____________

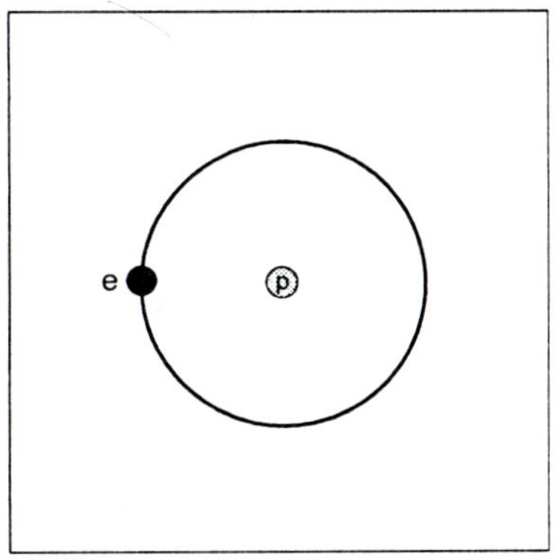

Figure C

Atomic Mass ____________

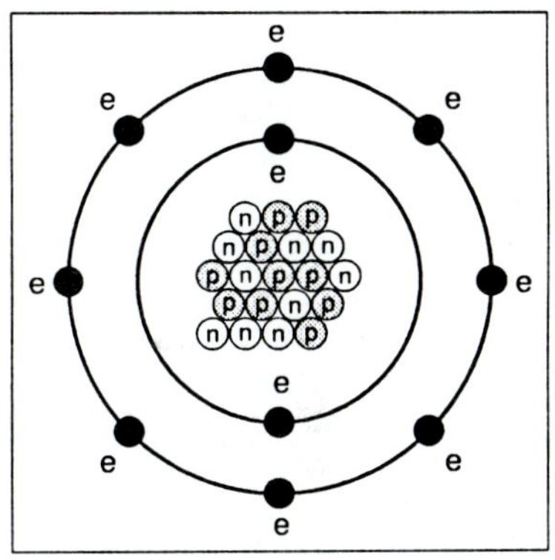

Figure D

Atomic Mass ____________

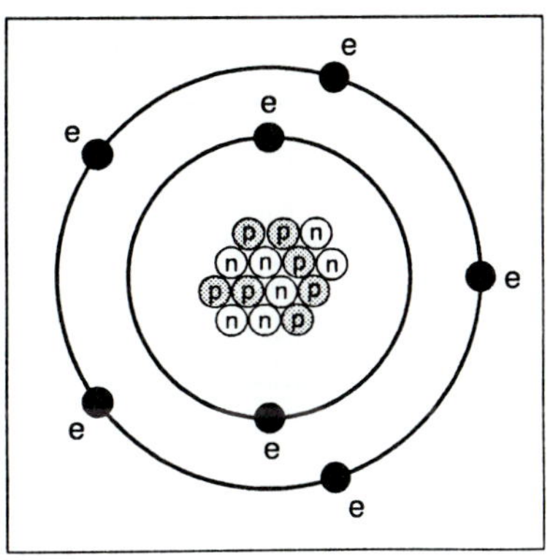

Figure E

Atomic Mass ____________

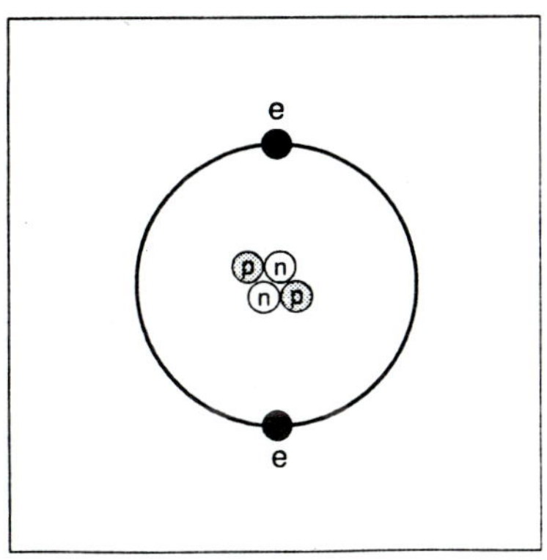

Figure F

Atomic Mass ____________

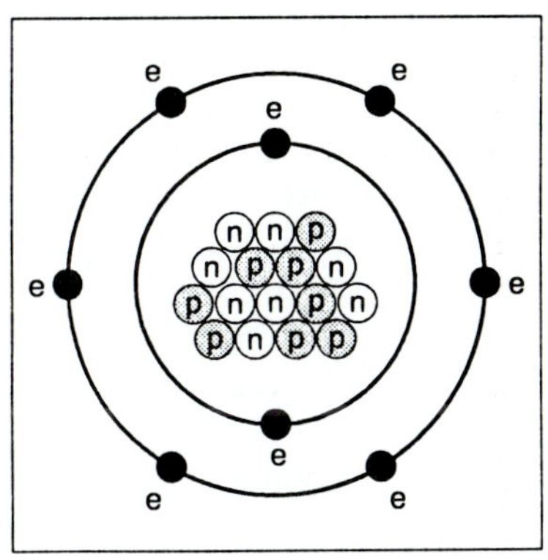

Figure G

Atomic Mass ____________

COMPLETE THE CHART

Complete the chart by filling in the missing information.

Name of element	Number of protons	Number of neutrons	Atomic mass
1. cobalt	27	32	
2. zinc	30	35	
3. krypton	36	48	
4. hydrogen	1	0	
5. potassium	19	20	
6. gold	79	118	
7. arsenic	33	42	
8. sulfur	16	16	
9. iodine	53	74	
10. tungsten	74	110	
11. silver	47	61	
12. uranium	92	146	
13. lead	82	125	
14. calcium	20	20	
15. oxygen	8	8	

WHAT DO THE PICTURES SHOW?

Each picture below shows an atom. Some information is given about each atom. Use this information to answer the questions about each atom.

REMEMBER, protons + neutrons = atomic mass

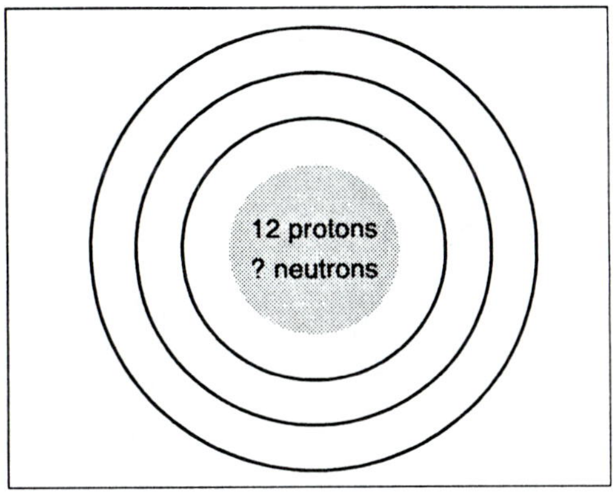

Figure H *Atomic mass = 24*

1. How many neutrons does this atom have? ____________

2. How many electrons? ____________

3. What is the atomic number? ____________

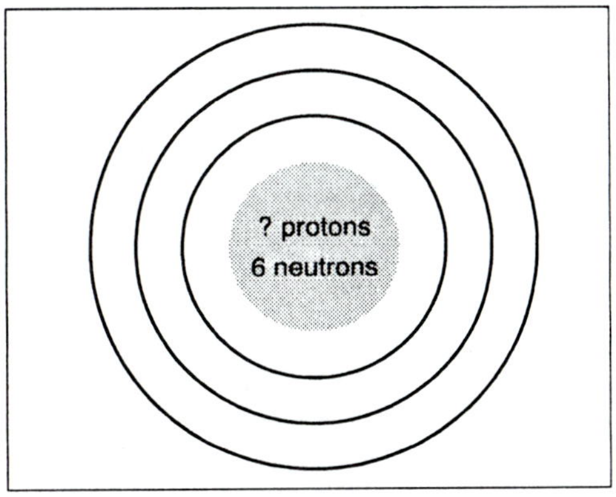

Figure I *Atomic mass = 11*

1. How many protons does this atom have? ____________

2. How many electrons? ____________

3. What is the atomic number? ____________

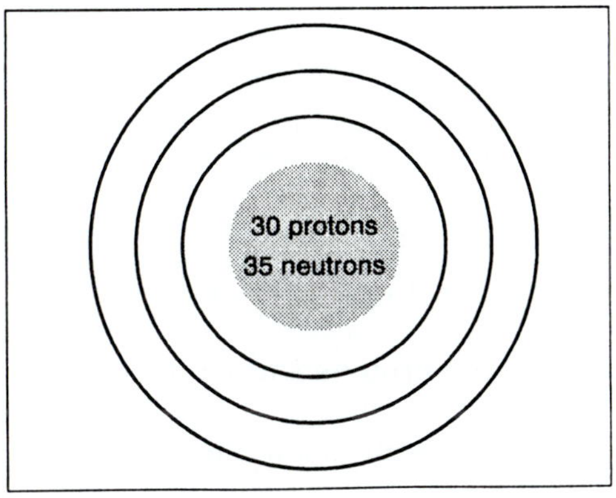

Figure J *Atomic mass = ?*

1. What is the atomic mass of this atom? ______

2. How many electrons? ____________

3. What is the atomic number? ____________

COMPLETE THE CHART

Complete the chart by filling in the missing information.

	Kind of Matter	Protons	Neutrons	Atomic Mass	Electrons	Atomic Number
1.	Oxygen	8		16	8	8
2.	Sodium			23	11	
3.	Carbon		6	12		
4.	Phosphorus		16			15
5.	Potassium	19	20			
6.	Iron	26		56		
7.	Copper	29	35	64		
8.	Chlorine			35		17
9.	Boron	5	6			
10.	Aluminum		14	27		

TRUE OR FALSE

In the space provided, write "true" if the sentence is true. Write "false" if the sentence is false.

__________ **1.** An atom has no mass.

__________ **2.** An electron is the largest part of an atom.

__________ **3.** All atoms have the same mass.

__________ **4.** All protons have the same mass.

__________ **5.** All oxygen atoms have the same mass.

__________ **6.** An oxygen atom has the same atomic number as a hydrogen atom.

__________ **7.** To find the atomic mass of an atom, we add the protons and electrons.

__________ **8.** The atomic number of an atom is the number of neutrons it has.

__________ **9.** Atoms of the same kind that have different numbers of neutrons are called isotopes.

__________ **10.** Atomic number = atomic mass.

What are elements?

13

element [EL-uh-munt]: simple substances than cannot be broken down into simpler substances
symbol: short way of writing the name of an element

LESSON 13 | What are elements?

There are 109 known <u>different</u> kinds of atoms that have been identified at the present time. This means that everything we know—every kind of matter—must be made up of one or more of these 109 kinds of atoms.

Most matter is made up of two or more kinds of atoms. But, some matter is made up of only one kind of atom. Any sample of matter that is made up of only one kind of atom is called an **element** (EL-uh-munt). An element is a simple substance that cannot be broken down into simpler substances.

Think about this. If there are only 109 kinds of atoms, there must be 109 different elements. There is one element for each kind of atom. The atoms of any one element are almost exactly alike. In addition, the atoms of any given element are different from the atoms of every other element.

You know the names of some elements—gold, copper, iron, oxygen, iodine, and probably many more. Only 92 of the known elements are found in nature. The other 17 have been made by scientists in the laboratory. Most of these have been made in very small amounts. In the future, more elements will probably be created. So, the total number of known elements will increase.

Most elements are found in the solid state. Some are found as gases. Only two elements—mercury and bromine—are found as liquids at room temperature.

The names of the elements can be written in **symbol** form. Symbols are signs that stand for something else. Here are some symbols that you may know: $ % ¢ & #. What does each one stand for?

RULES FOR WRITING SYMBOLS

1. The symbol always begins with a capital letter.
2. If there is a second letter, it is a lower case letter.
3. No period is used at the end of the symbol (unless it comes at the end of a sentence).

The symbols for the elements are the same all over the world. It makes no difference what country you are in—or what language is spoken there. The "language" of chemistry is the same all over the world.

MORE ABOUT ATOMS

Look at each figure below. Read about the figure. Then try to answer the questions.

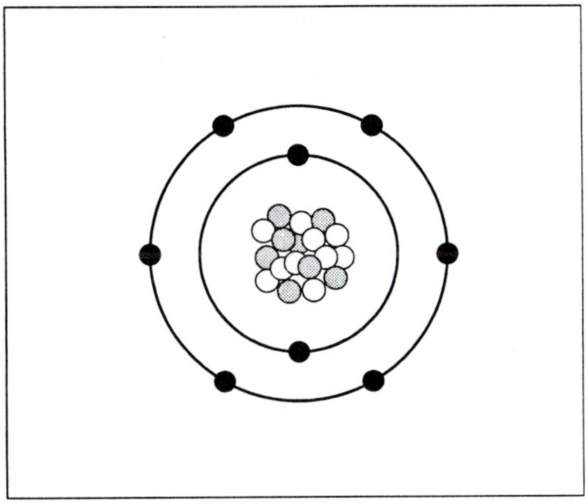

Figure A

This is a model of an atom.

Different elements have different atoms.

Sometimes the differences are very slight.

1. How many different kinds of atoms are there? ______________________

2. What do we call matter that has only one kind of atom? ______________________

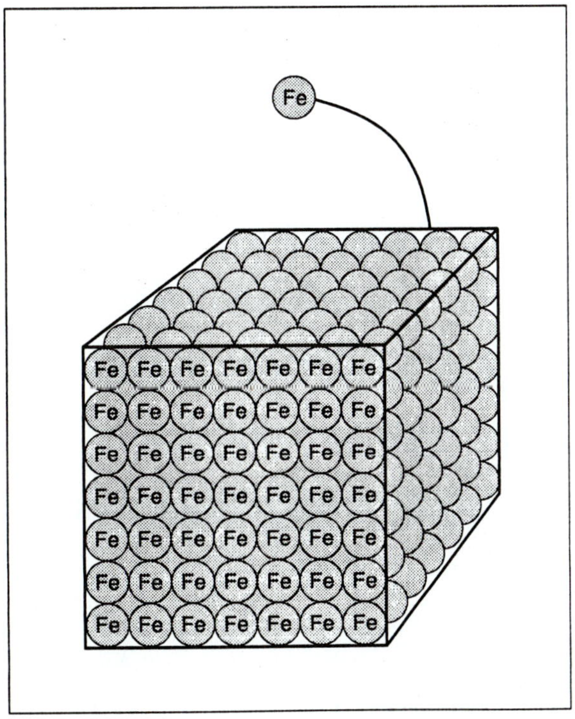

Figure B

The atoms of any one element are almost exactly alike.

Each of the circles in the picture stands for one atom of the element iron.

If we could study just one of these atoms, it would still have the properties of iron.

The smallest part of an element is just one atom of that element.

3. Is it possible to study just one atom?

____

4. Explain your answer. ______________________

____

____

____

FILL IN THE BLANK

Complete each statement using a term or terms from the list below. Write your answers in the spaces provided. Some answers may be used more than once.

different	one atom	109
element	92	alike
17	atoms	solid

1. All matter is made up of tiny particles called ________________ .
2. There are ________________ different kinds of atoms.
3. Matter that is made up of only one kind of atom is called an ________________ .
4. There are ________________ different elements.
5. There are ________________ natural elements.
6. Scientists have made ________________ elements in the laboratory.
7. The atoms of any one element are almost exactly ________________ .
8. The smallest part of an element is ________________ of that element.
9. Atoms of different elements are always ________________ .
10. Most elements are found in the ________________ state.

TRUE OR FALSE

In the space provided, write "true" if the sentence is true. Write "false" if the sentence is false.

__________ 1. Every atom in the world is exactly alike.

__________ 2. There are only 109 different kind of atoms known.

__________ 3. There are 90 different elements.

__________ 4. An element has only one kind of atom.

__________ 5. Atoms of different elements are alike.

__________ 6. Scientist have made 92 elements in the laboratory.

__________ 7. The symbols for elements are different in different countries.

__________ 8. An atom of iron is the same as an atom of oxygen.

__________ 9. The smallest part of an element is one atom.

__________ 10. Scientists may make new elements in the future.

ELEMENTS AND THEIR SYMBOLS

HOW DO THE ELEMENTS GET THEIR SYMBOLS?

The symbol of an element is taken from its name. Often the first letter of an elements's name is the symbol for that element. For example, the symbol for carbon is C.

Element	*Symbol*
oxygen	O
carbon	C
nitrogen	N
hydrogen	H

The name of an element may begin with a letter that is already the symbol for another element. In this case we use:

- the first two letters, as Os for osmium, or
- the first letter and some other letter from the name, as Cl for Chlorine.

Element	*Symbol*
osmium	Os
chlorine	Cl
calcium	Ca
helium	He
aluminum	Al
nickel	Ni

HOW DO ELEMENTS GET THEIR NAMES?

The symbols for most elements are taken from their names in other languages—mostly Latin and Greek. For example, the Latin name for iron is *ferrum*. The symbol for iron is Fe.

Element	*Symbol*	*Foreign Name*
lead	Pb	*plumbum* (Latin)
gold	Au	*aurium* (Latin)
barium	Ba	*barys* (Greek)
lithium	Li	*lithos* (Greek)

Several elements are named after places or after famous scientists. For example, Am for Americium and Es for Einsteinium.

Element	*Symbol*	*Named for*
californium	Cf	California
europium	Eu	Europe
curium	Cm	Marie Curie, French scientist
fermium	Fm	Enrico Fermi, Italian physicist

FILL IN THE BLANK

Complete each statement using a term or terms from the list below. Write your answers in the spaces provided.

capital	two	symbol
Greek	lower case	

1. The short way of writing an element is called its ______________ .
2. The first letter of a chemical symbol is always a ______________ letter.
3. If a chemical symbol has two letters, the second letter is always a ______________ letter.
4. A chemical symbol never has more than ______________ letters.
5. The symbols for many element come from Latin and ______________ words for the element.

MATCHING

Match each term in Column A with its description in Column B. Write the correct letter in the space provided.

	Column A	Column B
________	**1.** symbol	**a)** never used as a chemical symbol
________	**2.** each element	**b)** always a capital letter
________	**3.** second letter of a chemical symbol	**c)** has its own chemical symbol
________	**4.** three letters	**d)** sign that stands for something else
________	**5.** first letter of a chemical symbol	**e)** always a lower case letter

REACHING OUT

Why is it so important to have only one set of chemical symbols in the world?

____

____

____

____

What is the periodic table of elements?

14

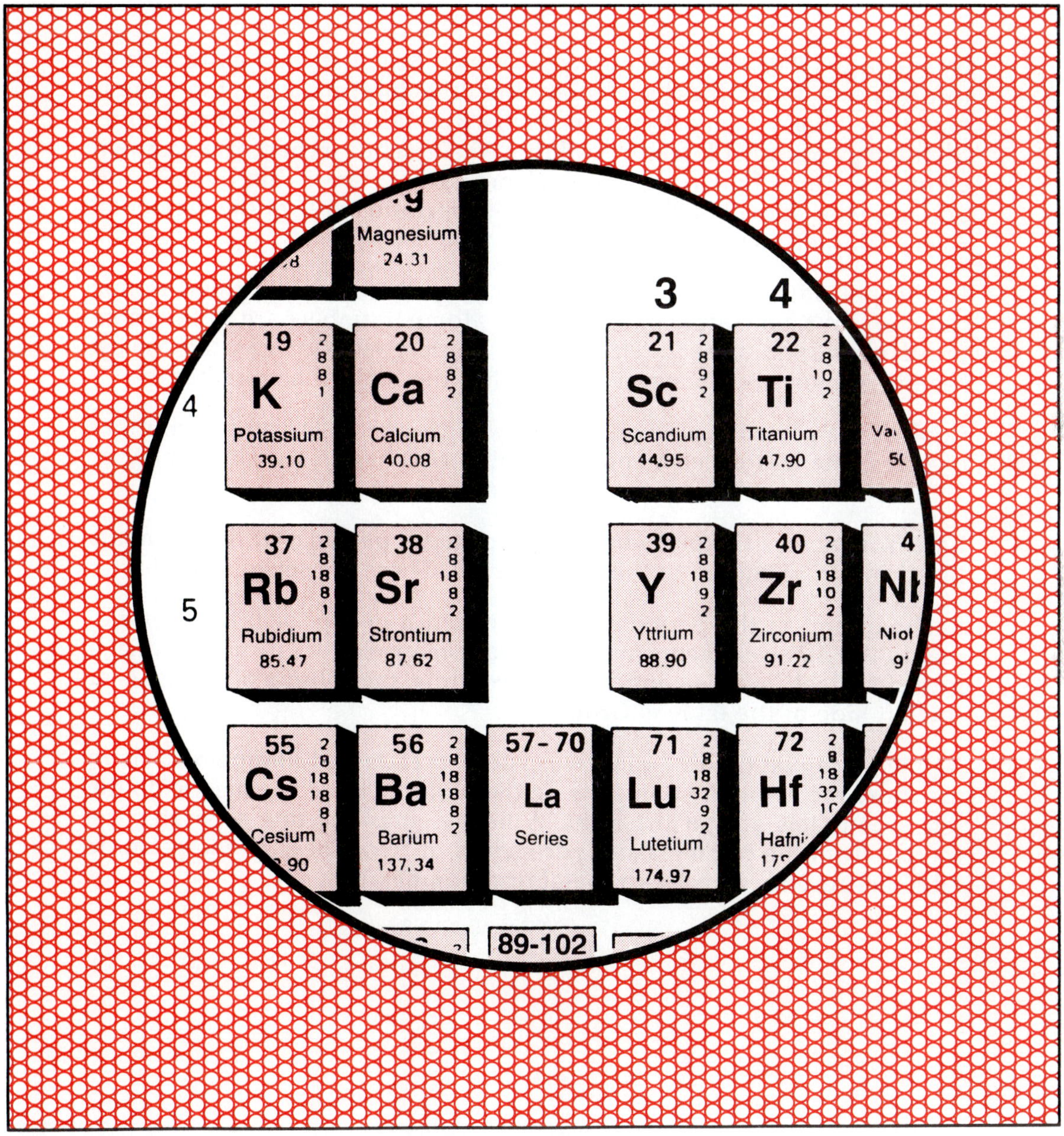

conductor: substance through which heat and electricity travel well
group: a column, or family, in the periodic table
metals: elements found on the left side of the periodic table, which share properties
nonmetals: elements found on the right side of the periodic table
period: a row across the periodic table
periodic [pir-ee-AHD-ik] **table:** a chart of all known elements

LESSON 14 What is the periodic table of elements?

By the mid-1800s, 60 elements were known. Scientists had quite a bit of information about these elements. But, the information wasn't organized, so it wasn't very useful. In 1869, a Russian scientist named Dmitri Mendeleev [duh-MEE-tree men-duh-LAY-uf] made a chart of the known elements. Since that time, more elements have been discovered and added to the chart. The chart is called the **periodic table of the elements**. Every country uses the same periodic table.

In the periodic table, elements are arranged in order of their atomic numbers. With a couple of exceptions, atomic numbers are in the same order as atomic masses. That is, the lightest element has the lowest atomic number; the heaviest element has the highest atomic number.

Look at the periodic table on page 81. Hydrogen has an atomic number of 1. It is the lightest element. Aluminum (Al) has an atomic number of 13. Only 12 elements are lighter than aluminum.

Each row across the periodic table is called a **period**. All the elements listed in a row belong to the same period. There are seven periods.

Each column in the periodic table is called a **group**, or family, of elements. All of the elements in a group have many similar properties. Each group is identified by a number. For example, the column of elements at the left side of the table is Group 1.

Elements can be divided into two types—**metals** and **nonmetals**. There are more metals than nonmetals. In the periodic table, metals are on the left, nonmetals are on the right. A heavy, step-like line separates the metals from the nonmetals. Hydrogen is in two places in the periodic table because it can act as a metal or a nonmetal.

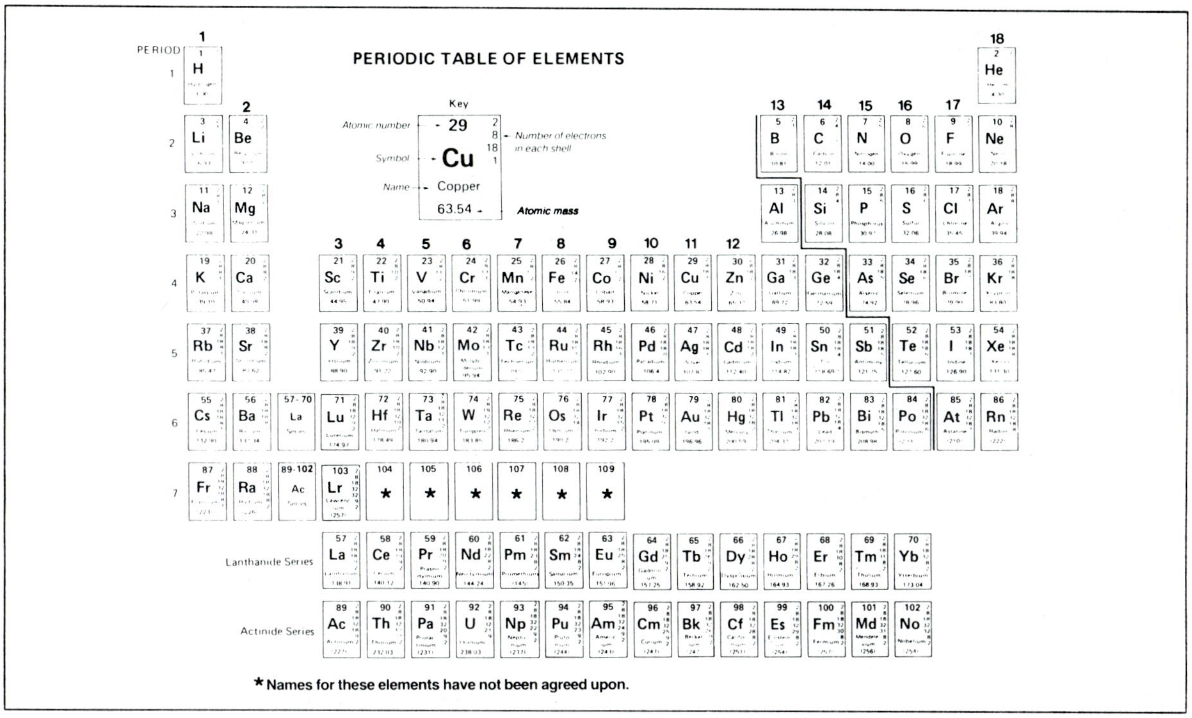

Figure A

USING THE PERIODIC TABLE

Answer the questions about the Periodic Table. The complete Periodic Table at the end of the book will tell you the names of the elements.

1. List the periods. ______________________________

2. List the groups. ______________________________

3. List the name, symbol, and atomic number of each element in Period 3.

Name	Symbol	Atomic Number

4. List the name, symbol, and atomic number of each element in Group 13.

Name	Symbol	Atomic Number

Finding the Element

The periods and families of five elements are listed below. Find each element in the Periodic Table. Then fill in the missing information.

	Period	Group	Elements	Symbol	Atomic Number
1.	2	16			
2.	3	1			
3.	4	8			
4.	1	18			
5.	6	12			

After you have completed the chart, answer the following questions about the elements on the chart.

a) Which one of these elements is the lightest? ______________________

b) You know it is the lightest because it has the ______________________ atomic number.
highest, lowest

Finding the Periods and Groups

The names of five elements are listed below. Find each element in the Periodic Table. Then fill in the missing information.

Period	Group	Element	Symbol	Atomic Number
		chlorine		
		potassium		
		neon		
		tin		
		krypton		

After you have completed the chart, answer the following questions about the elements on the chart.

1. Which of these elements are metals? ______________________

2. Which of these elements are nonmetals? ______________________

3. Two of these elements have many properties that are alike.

a) Name these elements. ______________________

b) We know they are alike because they are in the same ______________________ .
period, group

4. Name the metals that have properties like tin. ______________________

____

5. Name four elements that have properties similar to chlorine. ______________________

____

FILL IN THE BLANK

Complete each statement using a term or terms from the list below. Write your answers in the spaces provided. Some answers may be used more than once.

1869	group	period	sixty	higher
Mendeleev	Russia	mass	the same	atomic number

1. The Periodic Table was put together by a man named ______________ in the year ______________.
2. The man who put together the Periodic Table came from ______________ .
3. The number of known elements in 1869 was ______________ .
4. The elements are listed according to ______________ .
5. Each element is given a number called its ______________ .
6. An atomic mass has to do with the ______________ of an element
7. A heavy element usually has a ______________ atomic number than a light element does.
8. Elements in the same row across belong to the same ______________ .
9. Elements in the same up-and-down column belong to the same ______________ .
10. Elements of the same group have many properties that are ______________ .

COMPLETING SENTENCES

Choose the correct word or term for each statement. Write your choice in the spaces provided.

1. Elements in the same row across belong to the same ______________ .
   period, group
2. Elements in the same column down are members of the same ______________ .
   period, group
3. Elements that have many similar properties belong to the same ______________ .
   period, group
4. On the Periodic Table, metals are listed on the ______________ .
   right, left
5. On the Periodic Table, nonmetals are listed on the ______________ .
   right, left
6. There are more ______________ than ______________ .
   metals, nonmetals / metals, nonmetals
7. On the Periodic Table, elements are listed according to ______________ .
   alphabet, atomic number

What are the properties of metals?

15

conductor: substance through which heat and electricity travel well
malleable [MAL-ee-uh-bul]: able to be hammered into different shapes
metallic luster: shine of a metal

LESSON 15 What are the properties of metals?

Scientists divide all elements into two types—metals and nonmetals. They do this because most metals have certain properties that are the same. Most nonmetals share other properties that are the same.

You probably know many metals—tin, iron, silver, gold, copper, nickel. You would probably know if something is not a metal. But how would you explain what a metal is? What are the properties of metals? The properties of metals are:

- Metals are solid at room temperature. Think about iron, gold, copper, aluminum. They are all solids. Only one metal is not a solid at room temperature. That is mercury—a liquid.
- Metals are silver-gray. That is the color of all metal elements except two—gold and copper. Gold and copper are not silver-gray.
- Metals are good **conductors** of heat and electricity. That means that heat and electricity travel well through a metal.
- Metals have a certain shiny look. That shine is called **metallic luster**.
- Most metals are **malleable** [MAL-ee-uh-bul]. This means that you can hammer or press a piece of metal into a different shape without the metal breaking.

Metals have many uses because of their properties. For example, bridges, automobiles, ships, pots and pans, food cans, appliances, and jewelry are all made mostly of metal. How many other uses of metals can you think of?

CHOOSE THE RIGHT CAPTION

Below are six figures and six captions. Each caption matches one of the figures. Choose the caption that best describes each figure. Write the correct caption on the line provided.

Choose from these captions:

Metals are malleable.

Metals are solid except for mercury.

Metals are good conductors of heat.

Metals have a shiny metallic luster.

Metals are good conductors of electricity.

Metals are silver-gray—except gold and copper.

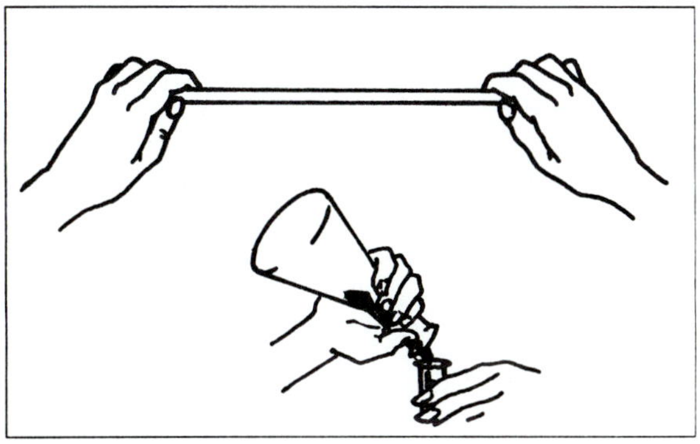

Figure A

____

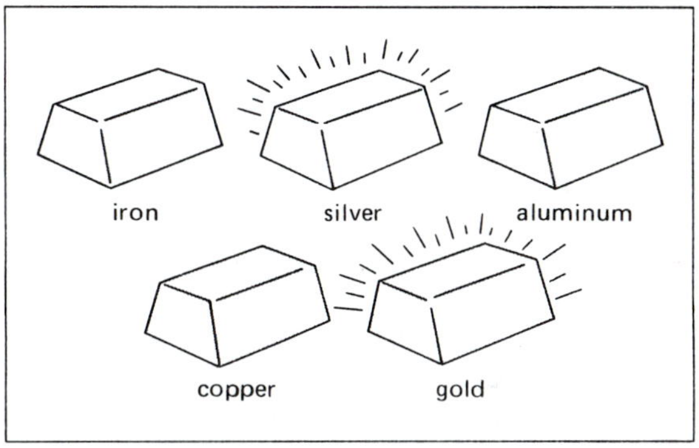

Figure B

____

Figure C

____

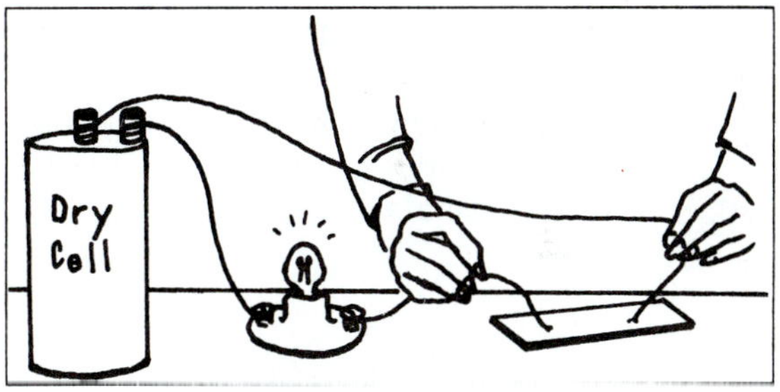

Figure D

____

Figure E

____

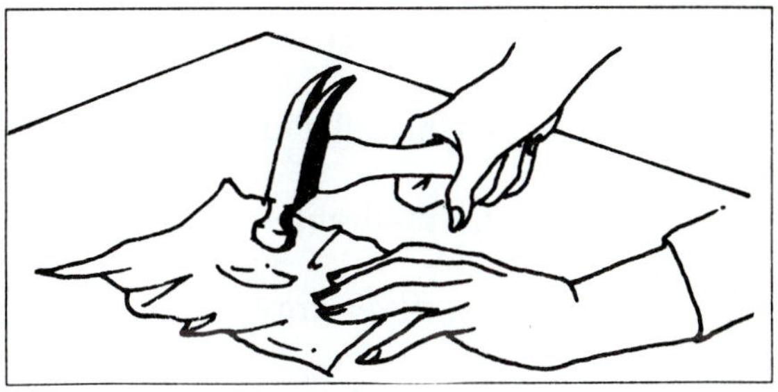

Figure F

____

FILL IN THE BLANK

Complete each statement using a term or terms from the list below. Write your answers in the spaces provided.

element	electricity	metallic luster
heat	mercury	malleable
metal	atoms	nonmetal
silver-gray	solid	gold
copper		

1. Matter is made up of tiny parts called ________________ .
2. Matter that has only one kind or atom is called an ________________ .
3. An element is identified as solid, liquid, or gas. An element is also identified as a ________________ or ________________ .
4. All metals except one are in the ________________ state.
5. The only metal that is a liquid at room temperature is ________________ .
6. The shine of a metal is called ________________ .
7. Most metals let some ________________ and ________________ pass through them.
8. Matter that can be hammered without breaking is said to be ________________ .
9. All but two metals have a ________________ color.
10. Only the metals ________________ and ________________ are not silver-gray.

MATCHING

Match each term in Column A with its description in Column B. Write the correct letter in the space provided.

	Column A	Column B
________	1. metallic luster	a) can be hammered without breaking
________	2. malleable	b) a liquid metal
________	3. silver-gray	c) has only one kind of atom
________	4. mercury	d) shine
________	5. element	e) color of most metals

ABOUT SOME METALS

Figure G *Metals are used where great strength is needed.*

Figure H *Aluminum is a light-weight metal. Most airplane bodies are made of aluminum.*

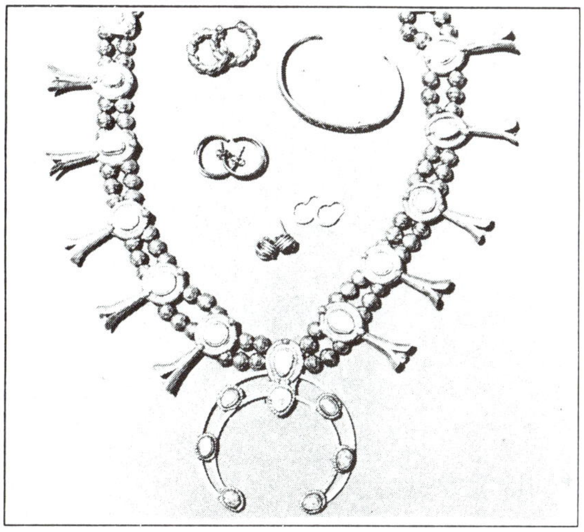

Figure I *Metals, like gold and silver, are used as jewelry.*

Figure J *Most electrical wires are made of the metal copper.*

WHICH PROPERTY?

Each of the following objects can be made because of a special property of metal. Which property makes it possible to make each of the objects? Write the name of the property in the space next to the object.

1. aluminum foil ______________________________
2. electrical wire ______________________________
3. pots and pans used for cooking ______________________________
4. automobile trim ______________________________
5. mercury thermometer ______________________________

TRUE OR FALSE

In the space provided, write "true" if the sentence is true. Write "false" if the sentence is false.

_________ 1. An element has only one kind of atom.

_________ 2. A metal is a kind of matter.

_________ 3. All metals are solid.

_________ 4. Metals conduct heat well.

_________ 5. Metals conduct electricity well.

_________ 6. Mercury is a metal.

_________ 7. Mercury takes the shape of its container.

_________ 8. Iron takes the shape of its container.

_________ 9. The color of copper is silver-gray.

_________ 10. Matter that is malleable breaks apart easily.

WORD SCRAMBLE

Below are several scrambled words you have used in this Lesson. Unscramble the words and write your answers in the spaces provided.

1. RUSTEL _________________
2. BLAMELEAL _________________
3. CRUMYER _________________
4. PROPEC _________________
5. LETAM _________________

REACHING OUT

Gold is a metal. So is aluminum. Why is gold so expensive and aluminum so cheap?

____

____

____

What are the properties of nonmetals?

16

insulator [IN-suh-layt-ur]: substance that does not conduct heat or electricity easily

LESSON 16 | What are the properties of nonmetals?

You know from your own experience what things are metals. You also have a good idea what things are not metals. What properties do all nonmetals have?

The properties of nonmetals are:

- Nonmetals are found in all states. Some are solids. Some are gases. And one nonmetal—bromine [BROH-meen]—is a liquid. All elements that are gases at room temperature are nonmetals.
- Nonmetals come in many colors. Bromine is a dark red liquid. Sulfur is yellow. Phosphorus [FAHS-fur-us] is white or red. Iodine is a steel-gray solid. Many gases have no color. They are invisible.

Some of the properties of nonmetals are exactly the opposite of the properties of metals.

- Nonmetals are poor conductors of heat and electricity. A substance that does not conduct heat and electricity is called an **insulator** [IN-suh-layt-ur].
- Nonmetals have no metallic luster.
- Nonmetals are not malleable. They break apart when they are hammered.
- Some nonmetals have odors. Chlorine is a greenish-yellow gas with a terrible odor. In fact, chlorine is poisonous.

We need nonmetals in our lives. The gases we breathe are nonmetals. Oxygen is the most important of these gases. Oxygen is a part of water, too. We cannot live without air and water.

WHAT DO THE PICTURES SHOW?

Look at each Figure. Then answer the questions.

Figure A

Figure B

Figure C

1. A malleable solid ________________ (does, does not) break apart when it is hammered.
2. The solid in this picture ________________ (is, is not) malleable.
3. The solid in this picture probably is a ________________ (metal, nonmetal) .
4. ________________ (Metals, Nonmetals) are malleable.
5. Sulfur is a ________________ (good, poor) conductor of electricity.
6. Sulfur is a ________________ (metal, nonmetal) .
7. Nonmetals are ________________ (good, poor) conductors of electricity.
8. Pots and pans are made of metals. Metals are ________________ (good, poor) conductors of heat.
9. The handle of the pan in this picture is covered with a ________________ (metal nonmetal) .
10. Nonmetals are ________________ (good, poor) conductors of heat.

MULTIPLE CHOICE

In the space provided, write the letter of the word that best completes each statement.

_______ 1. All nonmetals are

a) matter. b) energy.

c) liquids. d) malleable.

_______ 2. Matter that is a gas at room temperature must be a

a) conductor. b) metal.

c) nonmetal. d) insulator.

_______ 3. Nonmetals are

a) good conductors of heat and electricity. b) poor conductors of heat and electricity.

c) malleable. d) shiny.

_______ 4. We know that sulfur is not malleable because sulfur is

a) matter. b) a metal.

c) a nonmetal. d) solid.

_______ 5. An example of a poisonous gas is

a) oxygen. b) hydrogen.

c) bromine. d) chlorine.

TRUE OR FALSE

In the space provided, write "true" if the sentence is true. Write "false" if the sentence is false.

_______ 1. Nonmetals are found in solid, liquid, and gas states.

_______ 2. At room temperature, a metal can be invisible.

_______ 3. Some nonmetals are invisible.

_______ 4. Phosphorus is invisible.

_______ 5. Nonmetals have a metallic luster.

_______ 6. Nonmetals come in only a few colors.

_______ 7. Metals are malleable.

_______ 8. Nonmetals can be hammered. *(Careful, this is a trick question!)*

_______ 9. Nonmetals are good conductors of electricity.

_______ 10. Metals protect us from heat.

COMPLETE THE CHART

Complete the chart by filling in the missing information.

	Element	What is its state?	Is it a metal or nonmetal?	Does it have metallic luster?	Is it malleable?	Is it a good conductor of electricity?	Is it a good conductor of heat?
1.	aluminum						
2.	oxygen						
3.	phosphorus						
4.	iron						
5.	sulfur						
6.	chlorine						
7.	copper						

COMPLETE SENTENCES

Choose the correct word or term for each statement. Write your choice in the spaces provided.

1. Gases are ____________________ .
metals, nonmetals

2. This book is a ____________________ .
metal, nonmetal

3. Metals are ____________________ conductors of electricity.
good, poor

4. Nonmetals are ____________________ conductors of heat.
good, poor

5. Metals ____________________ malleable.
are, are not

6. ____________________ nonmetals are poisonous.
All, Some

7. Chlorine is a ____________________.
liquid, gas

8. Chlorine is ____________________ .
dangerous, pleasant smelling

9. ____________________ nonmetals have an odor.
Some, All

10. Oxygen is a nonmetal because it is ____________________ .
a gas, needed for life

SCIENCE *EXTRA*

Metallurgy

Look around and you probably will see a wide variety of metal objects. Metals have many uses in people's everyday lives. Metals are used to make cars, airplanes, electrical appliances, electrical wire, and many other objects. Where do the metals used to make all these things come from? Metals are found in the earth's crust. However, metals usually are found combined with other elements. A metallic ore is a rock or mineral from which a useful metal can be removed at a reasonable cost.

The science of removing metals from their ores, and preparing the metals for use is called metallurgy [MET-uh-lur-jee]. Metallurgy involves different steps and a variety of processes. The first step is to remove the metallic ore from the ground. Removal of the ore is carried out by mining. Large machines are used to take the ore from the ground.

After an ore is removed from the earth, waste materials that are combined with the ore are removed. This may be done by processes such as roasting. In roasting, sulfur and other impurities are removed from the ore by heating the ore in air. When heated, the sulfur combines with oxygen in the air and is released as a gas.

After waste materials are removed from an ore, the process of taking the desired metal from the ore begins. The most common method is called smelting. Smelting involves melting the ore to remove impurities. The melting usually takes place in a large furnace.

After a metal is taken from its ore, the metal is refined to make final products such as an aluminum can. The refining process may include combining the metal with another substance to give it special properties. For example, iron is the main element in steel. Iron is combined with nickel and chromium to make different kinds of steel.

Metallurgy is one of the oldest sciences. Some of the basic ideas of metallurgy have been used for many centuries. Ancient people knew how to remove iron from its ore as long as 4,000 years ago. However, modern metallurgy has methods and equipment that were only developed during the last fifty years.

What is energy?

17

energy: ability to make things move

LESSON 17 | What is energy?

Look around. How many things do you see moving?

What makes something move? Why does a rowboat drift downstream? Why do fallen leaves fly around? What lifts a rocket off of the ground? Why can you turn the page of this book?

Objects move because of energy. We say, therefore, that **energy** is the ability to make things move.

Energy is not like matter. Matter has mass and takes up space. Energy has <u>no</u> mass and does <u>not</u> take up any space.

There are seven forms of energy. They are:

- **CHEMICAL ENERGY** Chemical energy is what holds the atoms in molecules together. It can be released by chemical reactions like burning wood.
- **ELECTRICAL ENERGY** Electrical energy is the movement of electrons through matter. Electricity is a form of electrical energy.
- **HEAT ENERGY** Heat energy is the random motion, or vibration, of atoms within matter. The faster the atoms vibrate, the more heat energy they have.
- **LIGHT ENERGY** Light energy is the energy carried by light.
- **MECHANICAL ENERGY** Mechanical [muh-KAN-i-kul] energy is the energy of moving things.
- **NUCLEAR ENERGY** Nuclear [NEW-klee-ur] energy holds protons and neutrons together in the nucleus of atoms. This energy powers the sun and nuclear power plants.
- **SOUND ENERGY** Sound energy vibrates air molecules. The vibrating molecules move tiny bones in your ear. The message of sound then moves to your brain. This is how you hear. Sound energy can also vibrate objects.

Energy can be changed, or converted, from one form to another. For example, the chemical energy in gasoline is changed into heat energy in the engine of a car. This heat energy is then converted into mechanical energy to move the car.

FORMS OF ENERGY

Figure A *Mechanical energy*

Figure B *Electrical energy*

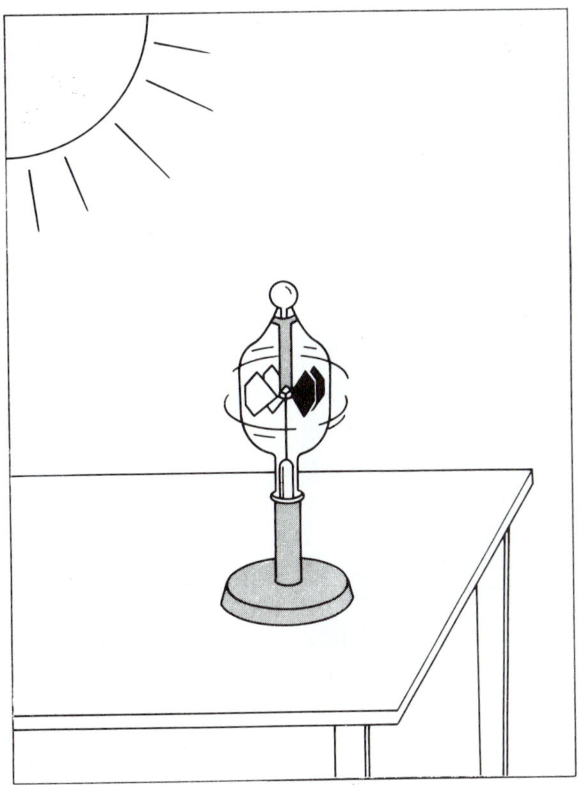

Figure C *Light energy*
A radiometer turns when light hits it.

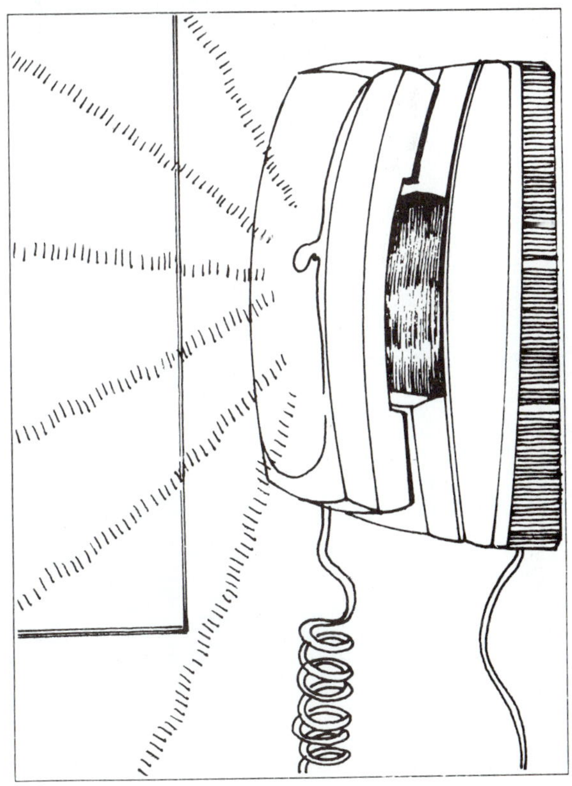

Figure D *Sound energy*

Figure E *Chemical energy*

Figure F *Heat energy*

Figure G *Nuclear energy*

FILL IN THE BLANK

Complete each statement using a term or terms from the list below. Write your answers in the spaces provided. Some words may be used more than once.

heat	chemical	
mechanical	energy	electrical
nuclear	sound	space
move	mass	light

1. The ability to make things move is called ________________ .
2. The main forms of energy are ________________ , ________________ , ________________ , ________________ , ________________ , ________________ , and ________________ .
3. The energy of moving things is called ________________ energy.
4. Moving things can make other things ________________ .
5. Thunder is an example of ________________ energy.
6. Matter has ________________ and takes up ________________ .
7. ________________ has no mass and does not take up space.

TRUE OR FALSE

In the space provided, write "true" if the sentence is true. Write "false" if the sentence is false.

________ 1. Energy makes things move.

________ 2. Only moving things have energy.

________ 3. There is only one form of energy.

________ 4. Everything has the same amount of energy.

________ 5. Living things need energy.

________ 6. Air molecules can vibrate.

________ 7. Electrical energy powers the sun.

________ 8. Nuclear energy powers the sun.

________ 9. Energy takes up space and has mass.

________ 10. Matter and energy are the same.

COMPLETING SENTENCES

Choose the correct word or term for each statement. Write your choice in the spaces provided.

1. ________________ has mass and takes up space.
   Matter, Energy

2. ________________ makes things move.
   Matter, Energy

3. Wind moving a windmill is an example of ________________ energy.
   atomic, mechanical

4. Humans have organs to detect ________________ energy.
   sound, electrical

5. Plants take light from the sun and store it as ________________ energy.
   chemical, atomic

WORD SCRAMBLE

Below are several scrambled words you have used in this Lesson. Unscramble the words and write your answers in the spaces provided.

1. GREYEN ________________
2. ONUDS ________________
3. TAMRET ________________
4. HITLG ________________
5. LAMCHINECA ________________

REACHING OUT

There is a limited amount of energy available on Earth. Why is it important that people conserve energy?

____

____

____

____

____

What are potential and kinetic energy?

18

kinetic [ki-NET-ik] **energy:** energy of motion
potential [puh-TEN-shul] **energy:** stored energy

LESSON 18 What are potential and kinetic energy?

A roller coaster speeds along its track. It has energy because it is moving. The energy of its motion is called kinetic [ki-NET-ik] energy. **Kinetic energy** is energy in action.

The roller coaster sits at the top of a hill. It is just about to speed down the track, but has not yet done so. It has energy because of where it is. This energy is called potential energy. **Potential energy** is stored energy that is not being used.

Substances like wood, coal, oil, and gasoline have stored energy because of their chemistry. They can burn. When the burn, they give off heat and light energy.

POTENTIAL ENERGY can be changed to KINETIC ENERGY.

Also, KINETIC ENERGY can be changed to POTENTIAL ENERGY.

The potential energy of the roller coaster when it is at the top of a hill is converted into kinetic energy as the roller coaster speeds down the hill. As the roller coaster goes up another hill, it slows down. The kinetic energy is converted into potential energy.

POTENTIAL OR KINETIC

Look at figures A-F. Write "potential energy" or "kinetic energy" on the correct lines.

Figure A

1. ______________________________

Figure B

2. ______________________________

Figure C

3. ______________________________

Figure D

4. ______________________________

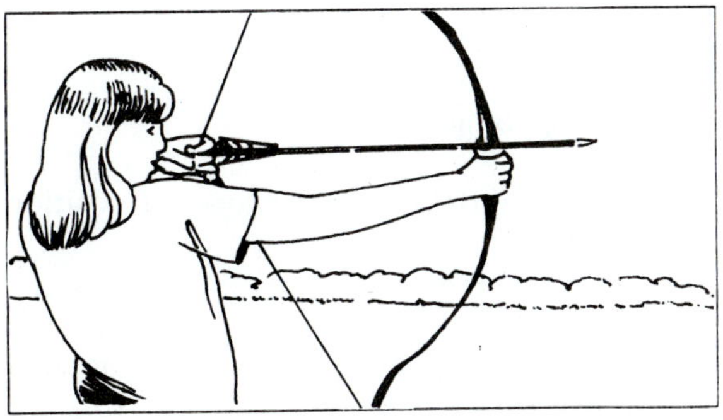

Figure E

5. ______________________________

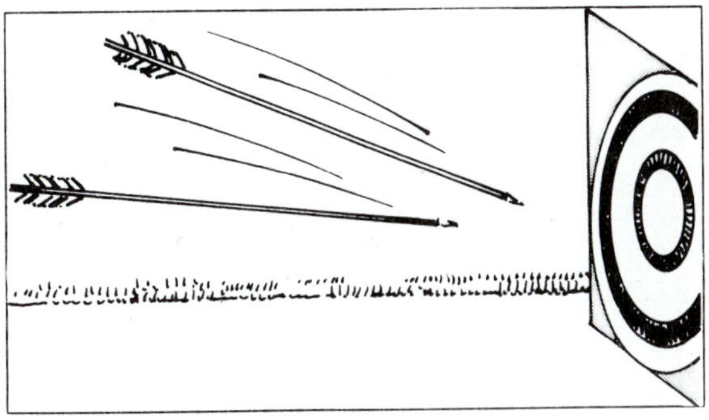

Figure F

6. ______________________________

FILL IN THE BLANK

Complete each statement using a term or terms from the list below. Write your answers in the spaces provided. Some words may be used more than once.

chemistry	piece of coal	kinetic
potential	energy	motion
burning coal	position	

1. The ability to make things move is called ________________ .
2. Stored energy is called ________________ energy.
3. Energy of motion is called ________________ energy.
4. Potential energy can change to ________________ energy.
5. A ________________ has potential energy.
6. ________________ has kinetic energy.
7. Matter has potential energy because of its ________________ or its ________________ .
8. A car at the top of a hill has potential energy because of its ________________ .
9. Fuel oil has potential energy because of its ________________ .
10. Matter has kinetic energy because of its ________________ .

MATCHING

Match each term in Column A with its description in Column B. Write the correct letter in the space provided.

	Column A	Column B
________	**1.** potential energy	**a)** energy in action
________	**2.** kinetic energy	**b)** has potential energy because of position
________	**3.** roller coaster	**c)** stored energy
________	**4.** change	**d)** has potential energy because of chemistry
________	**5.** oil	**e)** conversion

DIFFERENCES IN POTENTIAL ENERGY

Figure G

Look at the two rocks in Figure G. Rocks A and B are the same size and have the same mass. Rock B is higher than rock A. Therefore, rock B has more potential energy than rock A. The higher an object is, the more potential energy it has.

TRUE OR FALSE

In the space provided, write "true" if the sentence is true. Write "false" if the sentence is false.

__________ **1.** Objects that are higher have less potential energy.

__________ **2.** Objects that have kinetic energy do not move.

__________ **3.** Energy of position is potential energy.

__________ **4.** In Figure G, rock B has more potential energy.

__________ **5.** Potential energy and kinetic energy are the same.

__________ **6.** Potential energy can be changed into kinetic energy.

__________ **7.** A thrown baseball has kinetic energy.

__________ **8.** A match which has not been lit has potential energy.

__________ **9.** A lit match has potential energy.

__________ **10.** A car racing up a hill has kinetic energy.

CROSSWORD PUZZLE

Use the clues to complete the crossword puzzle.

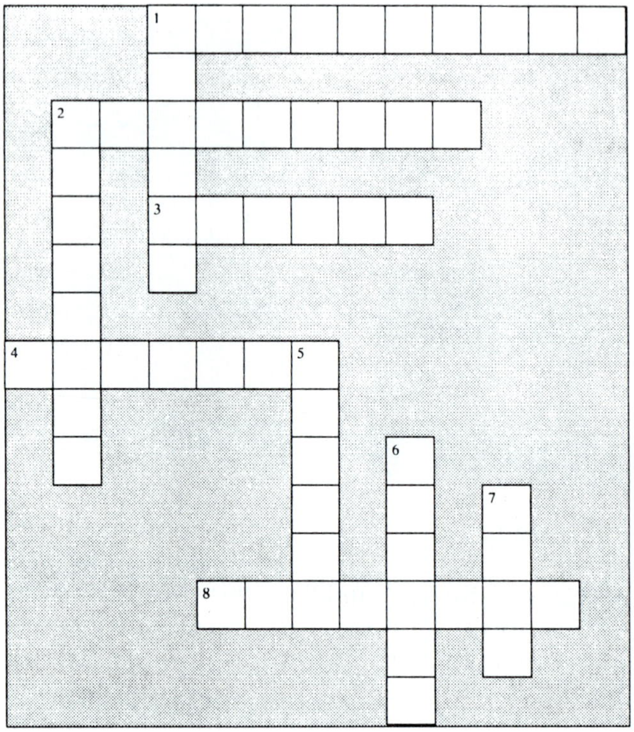

Clues

Across

1. energy of moving things
2. stored energy
3. ability to make things move
4. energy of action
8. energy that holds atoms in molecules together

Down

1. things that have mass and take up space
2. Potential energy is energy of ____ or chemistry.
5. conversion
6. movement
7. given off when burning wood

How does energy change form?

19

conversion [kun-VUR-zhunz]: energy change

Law of Conservation [kahn-sur-VAY-shun] **of Energy:** law which states that energy cannot be created or destroyed—it can only be changed

LESSON 19 How does energy change form?

Have you ever seen someone light a match? Before they light it, the match is cool. When it is lit, the match gives off light and heat. Where did this energy come from? There was chemical energy stored in the match. When the match was lit, the chemical energy changed into light and heat energy. We call energy changes **conversions** [kun-VUR-zhunz]. Conversion means a change into something else.

Energy conversions are very important. You cannot live without them.

The food you eat has chemical energy. Your body converts (changes) this chemical energy to the heat energy and the mechanical energy you need to stay alive.

Energy conversions are happening around you all the time. You flip up the switch in your home and the lights go on. Electrical energy has changed to light energy. Your toaster converts electrical energy to heat energy. A fan changes electrical energy to mechanical energy. Its motor gets hot. You can also hear it. This means that some of the energy has changed to heat and sound energy.

When energy is converted from one form to another, the amount of energy you start out with is the amount of energy you end up with. Only the forms of energy change.

We can put this idea in another way.

- Energy cannot be made.
- Energy cannot be destroyed. It is never "used up."
- Energy can just change from one form to another.

These three things are true for all kinds of energy. Scientists call this idea the **Law of Conservation** [kahn-sur-VAY-shun] **of Energy**. Conservation means saving. Although the energy changes its form, it is not lost.

ENERGY CONVERSIONS

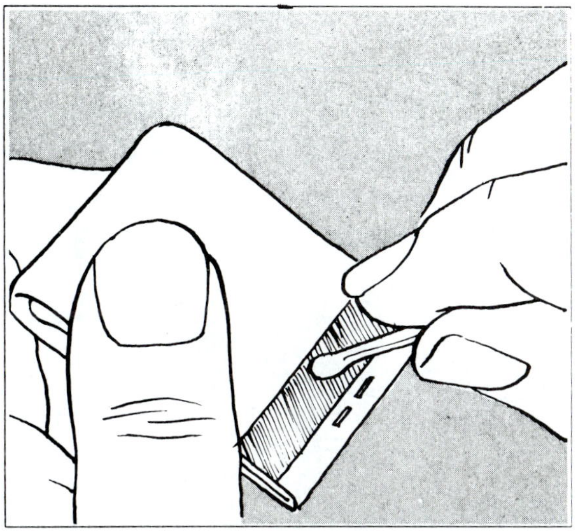

Figure A *Mechanical energy converts to heat energy.*

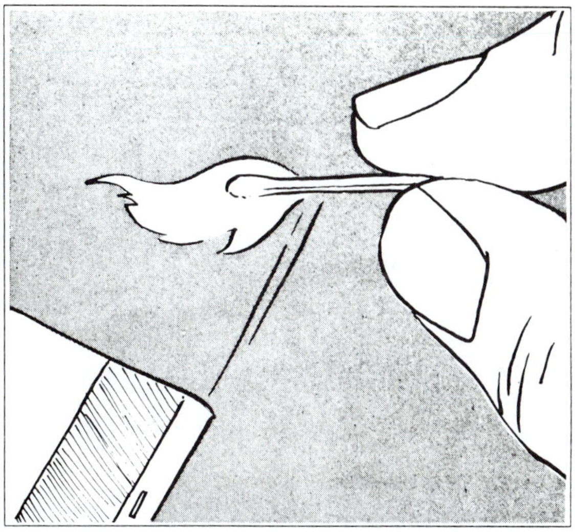

Figure B *Chemical energy converts to heat and light energy.*

ELECTRICAL ENERGY CONVERSIONS

Electrical energy can be converted to:

Figure C *Light energy*

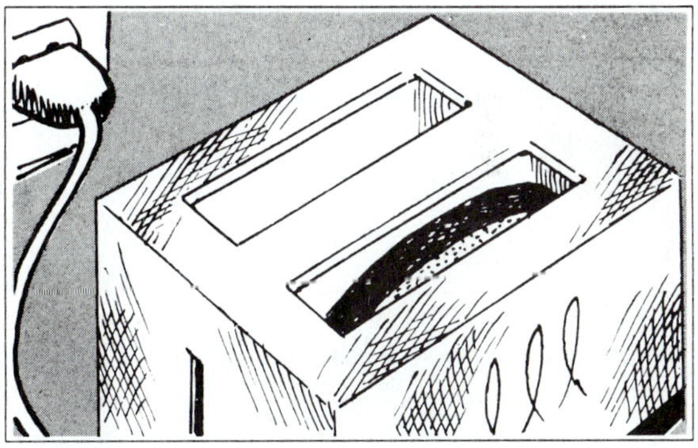

Figure D *Heat energy*

Figure E *Sound energy*

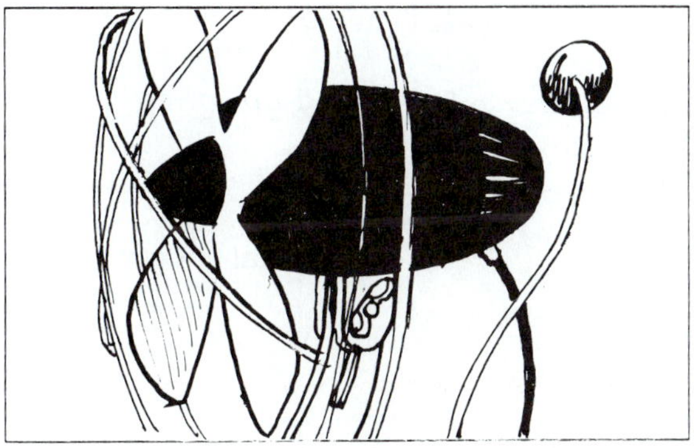

Figure F *Mechanical energy*

TRY THIS AT HOME!

Figure G

What You Need (Materials)

bowl
thermometer
eggbeater

What To Do (Procedure)

1. Fill a bowl with tap water.
2. Let it stand for several hours.
3. Take the temperature of the water.
4. Mix the water for two minutes with an eggbeater.
5. Take the temperature of the water again.

What You Saw (Observations)

1. The mixing made the water temperature ________________ .
   (rise/drop)

2. Mixing is an example of ________________ energy.

What You Learned (Conclusions)

3. This experiment shows that ________________ energy can be converted to ________________ energy.

MATCHING

Match each term in Column A with its description in Column B. Write the correct letter in the space provided.

	Column A	Column B
________	**1.** energy	**a)** has stored mechanical energy
________	**2.** wound-up spring	**b)** has stored chemical energy
________	**3.** conversions	**c)** energy of moving things
________	**4.** mechanical energy	**d)** the ability to make things move
________	**5.** battery	**e)** energy changes

WASTED ENERGY

Figure H

A car takes the chemical energy in gasoline and converts it into the mechanical energy of its motion. However, not all the chemical energy of the gasoline converts to mechanical energy. In fact, only a small amount of the chemical energy changes into energy that moves the car. The rest is waste. Some of the fuel's energy is wasted as sound. Most of the energy is wasted as heat.

Some waste energy is always given off when energy changes from one form to another. Most waste energy is in the form of heat.

People who design cars try to make them waste as little energy as possible. These people are called engineers. **Engineers** are people who take scientific ideas and put them to practical uses.

TRUE OR FALSE

In the space provided, write "true" if the sentence is true. Write "false" if the sentence is false.

__________ **1.** Heat is a form of energy.

__________ **2.** Heat is the only form of energy.

__________ **3.** Heat energy is sometimes useful.

__________ **4.** Heat energy is always useful.

__________ **5.** Energy can change from one form to another.

__________ **6.** Energy conversions always give off heat.

__________ **7.** Heat energy can change to mechanical energy.

__________ **8.** Mechanical energy can change to heat energy.

__________ **9.** Energy changes are called conservations.

__________ **10.** Energy changes are called conversions.

FILL IN THE BLANK

Complete each statement using a term or terms from the list below. Write your answers in the spaces provided. Some words may be used more than once.

body	mechanical	nuclear
electrical	conversions	chemical
change	sound	heat
light		move

1. Energy is the ability to make something ________________ .

2. The main forms of energy are ________________, ________________,

________________, ________________, ________________,

________________, and ________________.

3. The energy of moving objects is called ________________ energy.

4. Energy can ________________ from one form to another.

5. Energy changes are called ________________ .

6. Energy conversions take place inside your ________________ .

7. Food has ________________ energy.

8. The body changes chemical energy to ________________ and

________________ energy.

9. Motors sometimes give off wasted energy in the forms of ________________ energy and

________________ energy.

10. The useful energy a motor gives is ________________ energy.

WRITE IT DOWN

List the three parts of the "Law of Conservation of Energy." Start each part with the word "Energy."

1. ____

2. ____

3. ____

What is force?

20

gravity: force of attraction between all objects in the universe
inertia [in-UR-shuh]: resistance of an object to change its motion
Newtons: units used to measure force

LESSON 20 | What is force?

When you ride a bike, your foot pushes against the pedals. The push forces the wheels to turn.

When you drop something, it is pulled to the ground.

To turn a page, you first pull it up. Then you push it down. Try it. (At what point do you stop pulling and start pushing?)

Scientists call a push or a pull a force. A force is a push or a pull that changes the motion of an object.

Some forces are very weak. Others are very strong. For example, only a slight force will turn this page. More force is needed to pedal a bike. A much greater force is needed to lift heavy objects or to make a car go.

We see that a force can MAKE things move.

A force can do even more. A force can also

- STOP
- CHANGE THE SPEED, and
- CHANGE THE DIRECTION of something that is already moving.

A force can act in any direction. Forces are measured in units called **Newtons**. Newtons are named after Sir Isaac Newton, an English scientist who studied motion and gravity in the 1600s.

What happens to an object when no forces are acting on it? The object will continue to move at the same speed in the same direction. This characteristic of matter is called **inertia** [in-UR-shuh]. Inertia is the resistance of an object to change its motion. All matter has inertia.

PUSH OR PULL?

Each figure shows a push or a pull. Which one is it? Write the correct answers on spaces 1-4.

Figure A

1. ______________________

Figure B

2. ______________________

Figure C

3. ______________________

Figure D

4. ______________________

Figure E

When you drop a book, it falls to the ground. Why does this happen? The earth pulls on the book. The earth pulls on you. The earth pulls on everything. The measure of how much the earth pulls on you is called your weight.

All objects pull on one another. This pull is a force called **gravity**.

Every object has is own gravity. You have gravity. So does a piece of dust. Even a tiny atom has gravity. However, these pulls are too weak to be measured.

Usually, a larger, or more massive, object has a stronger gravity than a smaller, or less massive, object. The pull of the moon's gravity is less than the pull of the earth's gravity because the moon has less mass than the earth. Therefore, the moon's gravity is weaker. It pulls with less force. Since the moon's mass is one-sixth the mass of the earth, you would weigh about one-sixth as much on the moon as you would weigh on the earth.

Figure F

In outer space, you would be very far from both the earth and the moon. The pull of gravity would be zero. You would be weightless. Your weight would be zero. You would just float around.

FILL IN THE BLANK

Complete each statement using a term or terms from the list below. Write your answers in the spaces provided. Some answers may be used more than once.

stop	speed	any
move	direction	pull
push	gravity	less
energy		

1. The ability to make things move is called ______________ .
2. A force can ______________ something that is already moving.
3. A force is a ______________ or a ______________.
4. When you open a closet door, you ______________.
5. When you close a closet door, you ______________ .
6. A force can act in ______________ direction.
7. A force can make things ______________ .
8. A force can also make moving things change ______________ and ______________.
9. The earth's pull is called ______________ .
10. On the moon you would weigh ______________ than you do on the earth.

WORD SCRAMBLE

Below are several scrambled words you have used in this Lesson. Unscramble the words and write your answers in the spaces provided.

1. RAYGTIV ______________
2. ROCEF ______________
3. LUPL ______________
4. SUPH ______________
5. NOTNEW ______________

ACTION AND REACTION

Figure G *Space Shuttle Launch*

T-minus 10 seconds and counting to lift-off.

9 . . . 8 . . . 7 . . .

6 . . . 5 . . . 4 . . .

3 . . . 2 . . . 1 . . .

0

BLAST OFF!

The Space Shuttle is launched. Hot gases are forced out of the back of the Shuttle. In addition, these gases exert an equal and opposite force on the Shuttle. Every ACTION is always met with an equal and opposite REACTION.

You experience action and reaction forces every day. When you push against a table, the table pushes against you as well. If there were no reaction forces, rockets like the Space Shuttle would never work.

TRUE OR FALSE

In the space provided, write "true" if the sentence is true. Write "false" if the sentence is false.

_________ **1.** For every action there is a reaction.

_________ **2.** Actions and reactions work in the same direction.

_________ **3.** An object will slow down and stop as long as no forces act upon it.

_________ **4.** Inertia is a form of energy.

_________ **5.** Inertia is a property of matter.

_________ **6.** The gravitational pull of the moon is less than that of the earth.

_________ **7.** Pushing and pulling are two kinds of forces.

_________ **8.** Every object has its own gravity.

_________ **9.** A force can act in only one direction.

_________ **10.** Speeds are measured in units called Newtons.

UNDERSTANDING THE SPRING SCALE

How do we measure forces? Forces are measured with a spring scale. A force distorts (changes the shape of) a spring. A push squeezes it. A pull stretches it. The number next to the arrow or line on the scale tells us how strong the force is.

Fill in the blanks.

1. The force of gravity is a

   ______________ .
   push, pull

2. The force makes the spring move

   ______________ .
   up, down

3. The heavier the object,

   the ______________ the force.
   less, greater

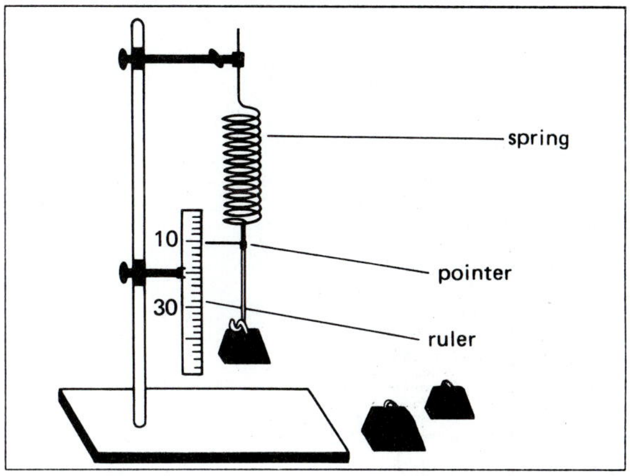

Figure H

4. The stronger the force, the ______________ the spring stretches.
   more, less

5. The weaker the force, the ______________ the spring stretches.
   more, less

6. A 1-gram mass exerts ______________ force than a 2-gram mass.
   more, less

7. A 2-gram mass stretches a spring ______________ as a 1-gram mass.
   twice as much, half as much

8. A 1-gram mass stretches a spring ______________ as a 2-gram mass.
   twice as much, half as much

9. Fill in the following chart.

Mass	Stretch
10 grams	1 cm
20 grams	
	3 cm
45 grams	
	5 cm

10. Using arithmetic and the chart above, figure out how much a kilogram would stretch the spring. ______________

MATCHING

Match each term in Column A with its description in Column B. Write the correct letter in the space provided.

	Column A	Column B
_______	**1.** Gravity	**a)** opposite of pull
_______	**2.** Force	**b)** opposite of action
_______	**3.** Inertia	**c)** measures forces
_______	**4.** Action	**d)** changes
_______	**5.** Reaction	**e)** causes objects in motion to remain in motion
_______	**6.** Push	**f)** opposite of reaction
_______	**7.** Pull	**g)** pull of the earth on you
_______	**8.** Spring scale	**h)** push or pull
_______	**9.** Conservation	**i)** energy can not be created or destroyed
_______	**10.** Conversion	**j)** opposite of push

REACHING OUT

A moving car has inertia which tends to keep it moving in a straight line. All of the passengers have inertia as well. If the car suddenly stops, what would happen to the passengers if they were not wearing seatbelts? Why is it a good idea to wear seatbelts?

____

____

____

____

____

____

____

____

____

What happens when forces work at the same time?

21

resultant: final amount and direction of a force
vector [VEK-tur]: an arrow that shows the strength and direction of a force

LESSON 21 What happens when forces work at the same time?

When you push a cart, only one force is at work on the cart—you. If your friend helps you, a second force is added. Both forces work together.

In arm wrestling, two forces push against each other.

More than one force can work on an object at the same time. Each force can push or pull in any direction. What happens when the forces act depends upon two things: their strength and their directions.

Let's study two simple examples—forces that work with each other and forces that work against each other.

FORCES THAT WORK AGAINST EACH OTHER Two things can happen when forces work in opposite directions.

- They can balance each other.
- One force can overpower the other force.

Opposing forces balance each other when they have the same strength. When forces are balanced, nothing moves.

A force is overpowered when a stronger force works against a weaker force. The object moves in the direction that the stronger force is pushing or pulling.

It is easy to figure the final force of opposite forces. Just subtract the weaker force from the stronger force.

A final force and its direction is called a **resultant** [ri-ZULT-unt].

FORCES THAT WORK WITH EACH OTHER Forces working together push or pull in the same direction. They help each other. The force becomes greater.

You find a resultant force of this kind by adding the helping forces.

Forces working at an angle also affect each other. But the resultant is not as easy to figure out.

WORKING WITH FORCES

The following figures show forces at work. Study each figure. Then answer the questions for each.

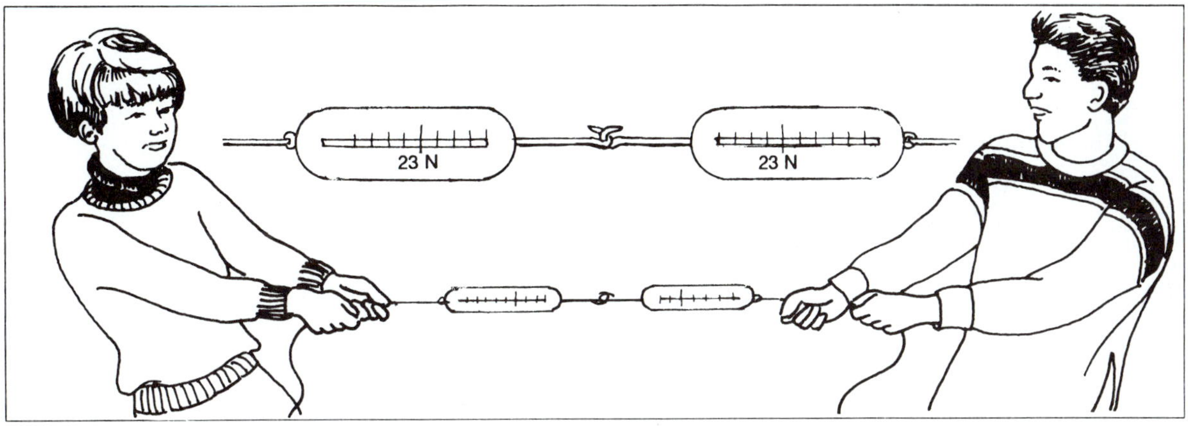

Figure A

1. The forces in Figure A are ______________________________ .
   pushing, pulling, pushing and pulling

2. The forces are ______________________________ .
   working together, working opposite to each other

3. The forces are ______________ .
   equal, not equal

4. The forces ______________ balance each other.
   do, do not

5. The resultant force is ______________________________ .
   23 Newtons to the right, 23 Newtons to the left, zero

6. There ______________ movement.
   is, is no

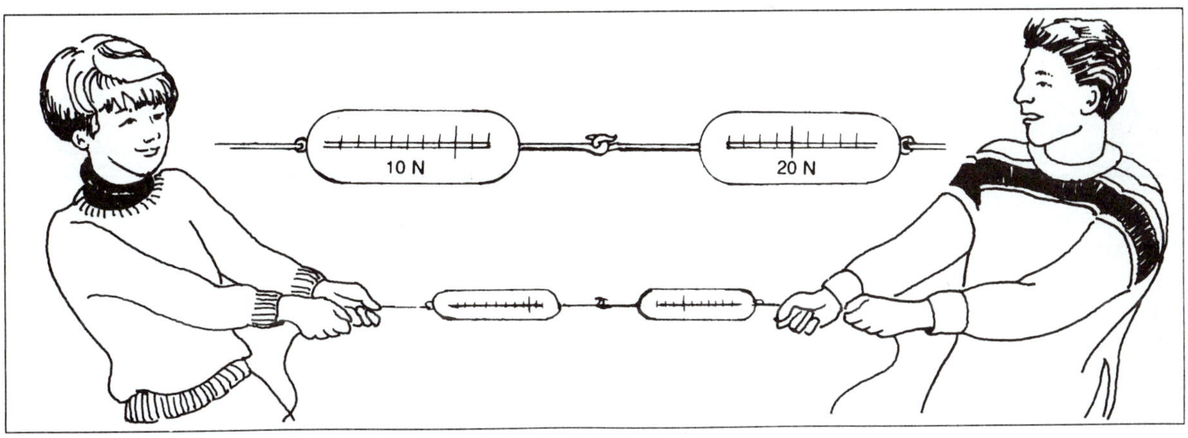

Figure B

7. The forces in Figure B are ______________________ .
pushing, pulling

8. The forces are ____ .
working together, working opposite each other

9. The forces ______________________ equal.
are, are not

10. The stronger force is pulling to the ______________________ .
right, left

11. The weaker force is pulling to the ______________________ .
right, left

12. There ______________________ movement.
is, is no

13. The ______________________ force is overpowering the ______________________ force.
stronger, weaker — stronger, weaker

14. The movement is to the ______________________ .
right, left

15. The resultant force is ____ .
20 Newtons to the right, 10 Newtons to the left, 10 Newtons to the right, zero

Figure C

16. The forces in Figure C are ______________________________________ .
pushing, pulling, pushing and pulling

17. The forces are ____ .
working together, working opposite each other

18. The forces ____________ equal.
are, are not

19. The stronger force is pushing to the ____________ .
right, left

20. The weaker force is pushing to the ____________ .
right, left

21. There ____________ movement.
is, is no

22. The ____________ force is overpowering the ____________ force.
stronger, weaker / stronger, weaker

23. The movement is to the ____________ .
right, left

24. What is the resultant? [Remember you need a force <u>and</u> a direction.]

____

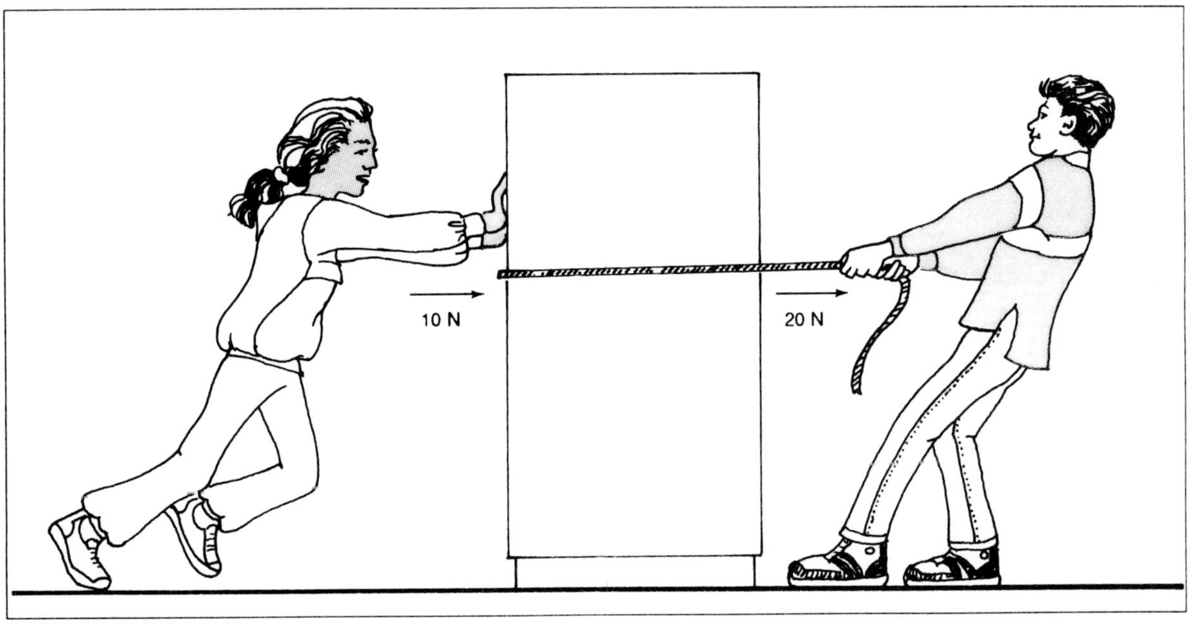

Figure D

25. The forces in Figure D are ____________ .
pushing, pulling, pushing and pulling

26. The forces are working ____________ .
together, opposite each other

27. a) Is there movement? ____________

b) In which direction: ____________

28. What is the resultant? ____________

FILL IN THE BLANK

Complete each statement using a term or terms from the list below. Write your answers in the spaces provided.

overpower	help	pull
stronger	push	resultant
against	subtract	is
is no	balance	add

1. A force is a ______________ or a ______________.
2. Opposite forces work ______________ each other.
3. Opposite forces can either ______________ or ______________ one another.
4. When forces are balanced, there ______________ movement.
5. When one force overpowers another force, there ______________ movement.
6. A final force and its direction is called a ______________ .
7. To find the resultant of opposite forces, you ______________ the forces.
8. Forces working together ______________ each other.
9. A helping force makes a force ______________ .
10. To find the resultant of forces working together, you ______________ the forces.

HOW DO YOU SHOW A FORCE?

You show a force with a **vector** [VEC-tur]. A vector is a line with an arrow. It starts with a dot.

- The dot shows where the force begins.
- The length shows the strength of the force.
- The arrow shows the direction.

Figure E shows an example of a vector.

In Figure E each box stands for a force of one Newton.

The vector in Figure E stands for a 5 Newtons force to the right.

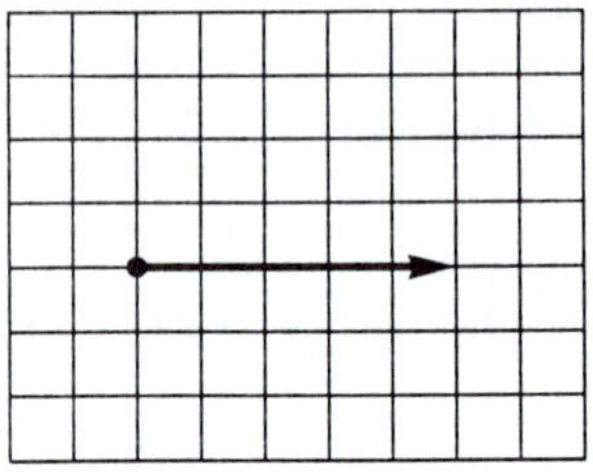

Figure E

READING VECTORS

Fill in the chart from the information in Figure F. Each box equals a force of one Newton.

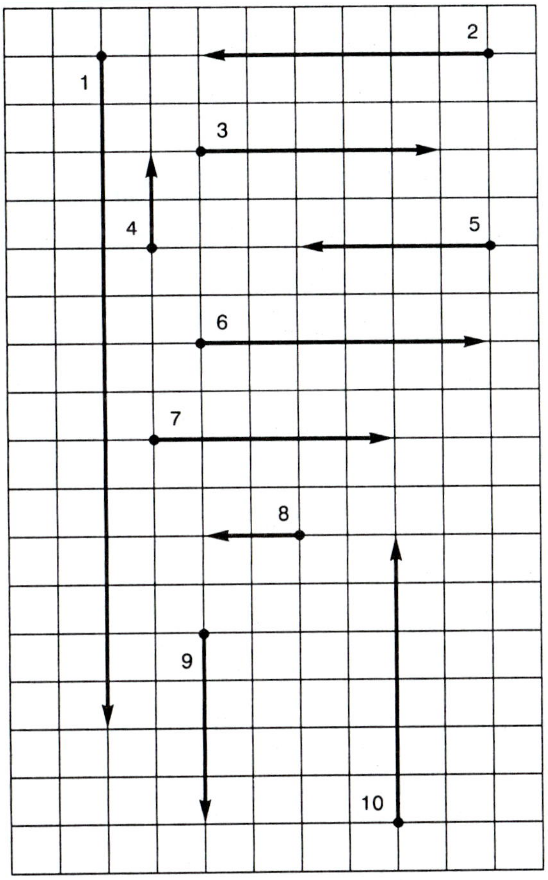

Figure F

	Force [Newtons]	Direction [right, left, up, down]
1.		
2.		
3.		
4.		
5.		
6.		
7.		
8.		
9.		
10.		

DRAWING VECTORS

Draw each vector in the space provided. Start at the dot. Each box stands for 1 Newton of force.

1. 8 N force to the right:

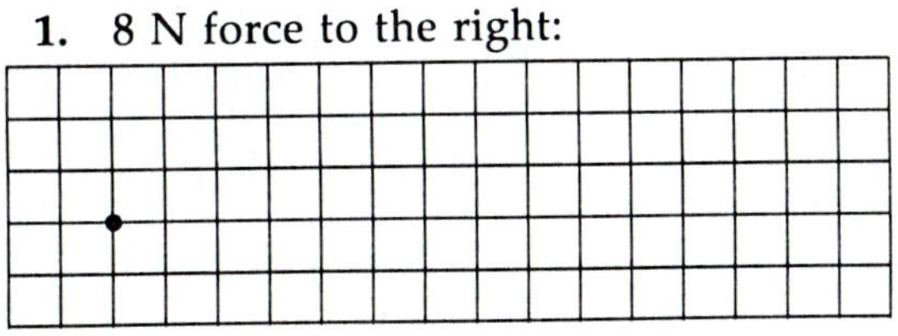

Figure G

2. 13 N force to the left:

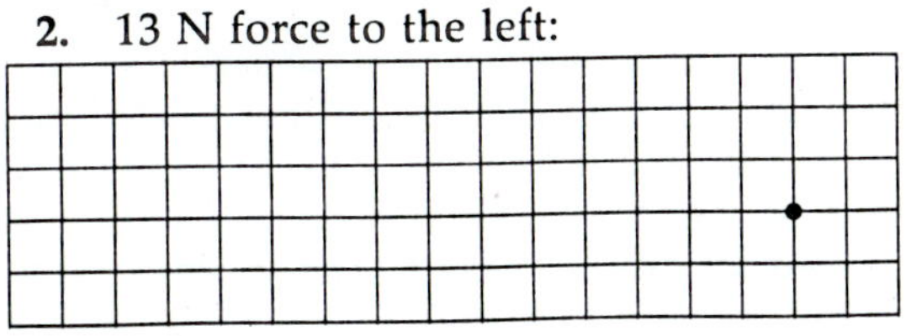

Figure H

3. 7 N downward force:

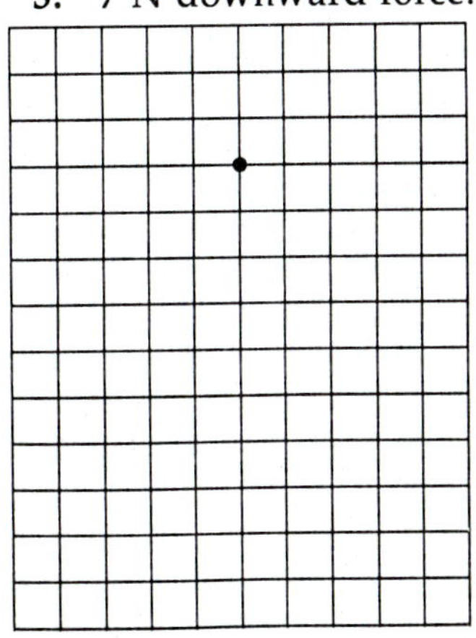

Figure I

FIGURING RESULTANTS

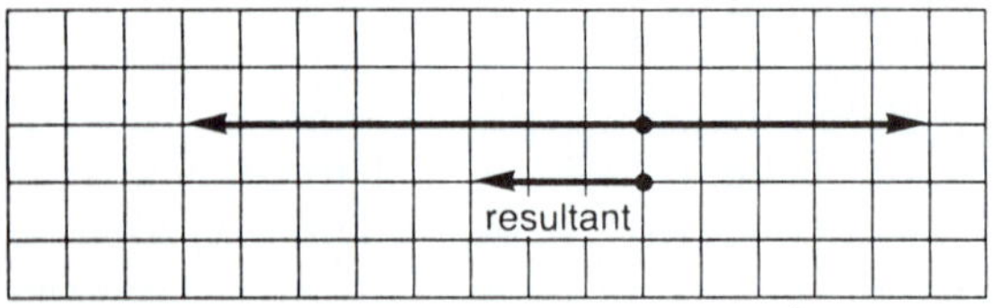

Figure J

Figure J shows two opposite forces. There is a 5 Newton force to the right and an 8 Newton force to the left. We subtract 5 from 8. The resultant is a 3 Newton force to the left. We show the resultant vector.

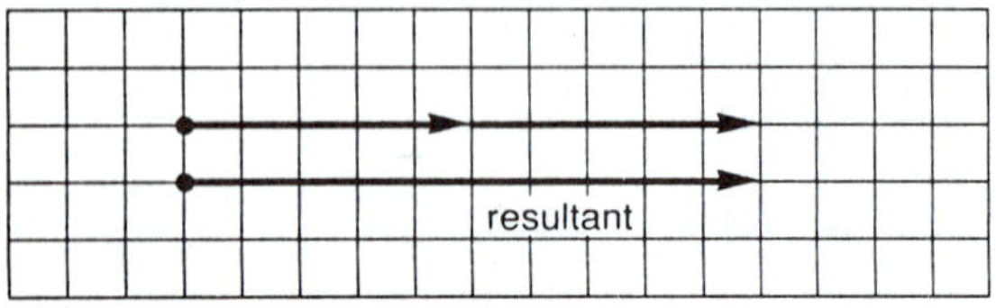

Figure K shows two helping forces. They are both 5 Newton forces to the right. We add 5 and 5. The resultant is a 10 Newton force to the right.

Figure K

DRAWING RESULTANTS

Six sets of vectors are shown below. Study each set. Draw the resultant vector next to each set. Start at the dot. The first one has been done for you. [Careful, one of the resultants is zero.]

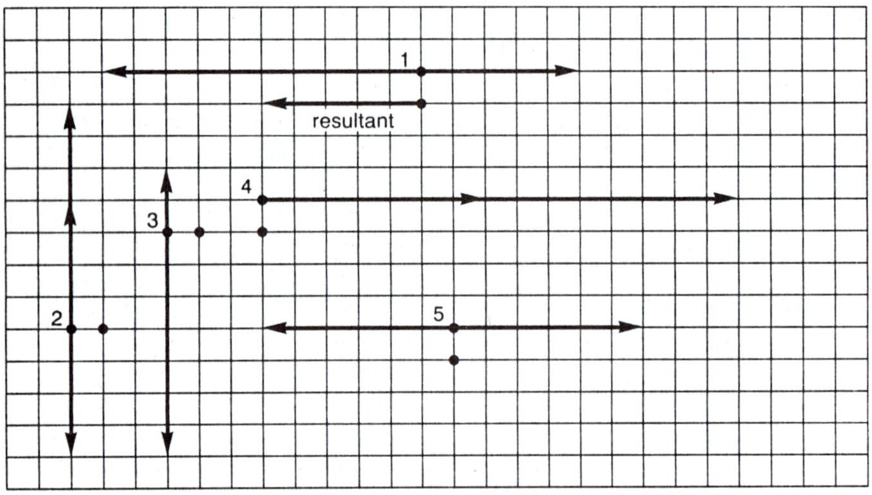

Figure L

Now fill in the chart from your work on Figure L. The first one has been done for you.

	Total Number of Forces	Amount of Force (Newtons)	Direction (right, left, up, down)	Resultant	Movement? (YES or NO)
1.	2	10 5	left right	5 left	YES
2.					
3.					
4.					
5.					

What is work?

22

Joules [JOOLS]: units used to measure work
work: a force that acts over a distance

LESSON 22 What is work?

A strong man pushes hard against a wall. He pushes and pushes with all his might. He huffs and he puffs. But no matter how hard he pushes, he cannot move the wall.

A small boy sees a leaf on his overalls. He flicks it away easily with one finger.

Here is a strange question: Who did more work, the strong man or the little boy?

The question may seem strange, but the answer may seem even stranger. The strong man who pushed against the wall did <u>no</u> work at all. The little boy who flicked away the leaf did <u>some</u> work. He did very little work, but it was still work.

What can explain such a strange answer?

Both the strong man and the little boy used force. But only the little boy's force made something move. The man could not make the wall move.

For **work** to be done, two things are needed:

- There must be a force.
- The force must make something move.

If there is no movement, no work is done—no matter how strong the force is.

There is a simple formula to help figure out how much <u>work</u> has been done:

Work = Force × Distance
W = F × D

Work is measured in units called **Joules** [JOOLS]. Since Force is measured in Newtons and distance is measured in meters, Joules are equal to Newtons times meters.

Joules = Newtons × Meters
J = N × m.

HOW MUCH WORK?

Figure A

This boy is pushing with a force of 15 Newtons. The rock does not move. How much work has the boy in Figure A done?

The formula for work is:

Work = Force × Distance

Work = 15 Newtons x 0 meters

Work = 0 Joules (any number multiplied by zero is zero)

Force without movement means no work is done.

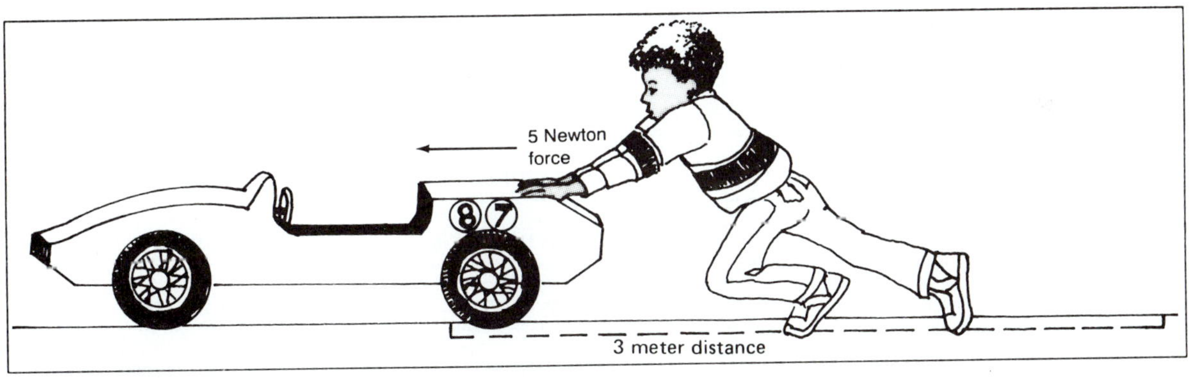

Figure B

This boy has pushed a cart 3 meters. He used a force of 5 Newtons.

How much work has the boy in Figure B done?

Work = Force × Distance

Work = 5 Newtons × 3 meters

Work = 15 Joules

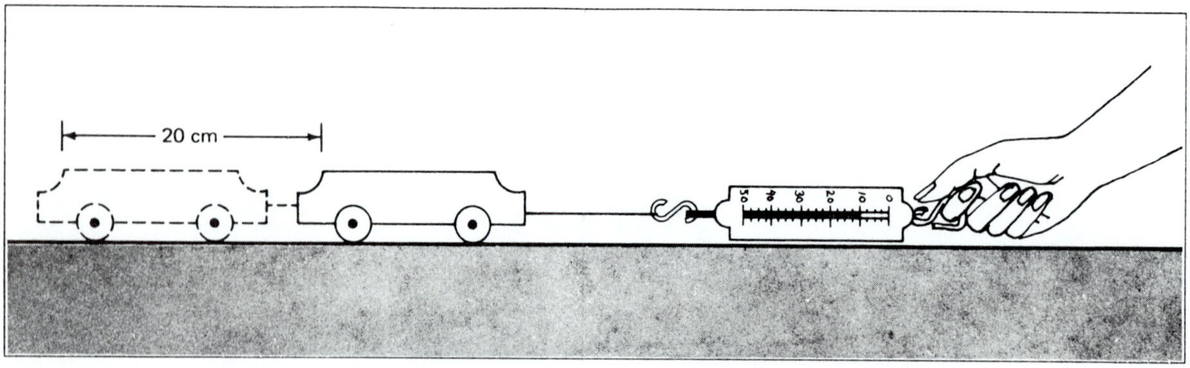

Figure C

This girl has pulled a small laboratory carriage 0.2 meters. She used a force of only 10 Newtons.

How much work has the girl in Figure C done?

Work = 10 Newtons × 0.2 meters

Work = 2 Joules

FIGURING WORK

Figure the amount of work done in each example. Write your answer in the work column. Use the margin to do your arithmetic. The first problem has been done for you.

	Work	= Force	× Distance
1.	10 Joules	5 Newtons	2 meters
2.		5 Newtons	3 meters
3.		5 Newtons	6 meters
4.		6 Newtons	5 meters
5.		6 Newtons	6 meters
6.		6 Newtons	7 meters
7.		10 Newtons	5 meters
8.		5 Newtons	20 meters
9.		50 Newtons	no distance
10.		200 Newtons	.6 meters
11.		100 Newtons	30 meters

Find the <u>distance</u> in these examples. The first has been done for you.

Work =	Force ×	Distance
20 Joules	2 Newtons	10 meters
none	10 Newtons	
100 Joules	1 Newton	
60 Joules	3 Newtons	
3 Joules	6 Newtons	
25 Joules	5 Newtons	

Find the <u>force</u> in these examples. The first has been done for you.

Work =	Force ×	Distance
30 Joules	2 Newtons	15 meters
150 Joules		10 meters
25 Joules		5 meters
120 Joules		6 meters
25 Joules		.1 meter
75 Joules		75 meters

MATCHING

Match each term in Column A with its description in Column B. Write the correct letter in the space provided.

	Column A	Column B
______	**1.** Newtons	**a)** measured in meters
______	**2.** Joules	**b)** measured in Newtons
______	**3.** Work	**c)** named after English Scientist who lived in the 1600s
______	**4.** Force	**d)** measured in Joules
______	**5.** Distance	**e)** Newton × meters

FILL IN THE BLANK

Complete each statement using a term or terms from the list below. Write your answers in the spaces provided.

weaker	zero	work
Newtons	Joules	$W = F \times D$
distance	force	stronger
equal	push	pull
opposite		

1. Work needs two things: ________________ and ________________ .
2. The formula for work is ________________ .
3. If there is no movement, there can be no ________________ .
4. Any number multiplied by zero is ________________ .
5. Work is measured in ________________ .
6. Forces working against each other create a ________________ force than if the two forces were working together.
7. Forces are measured in ________________ .
8. A force is a ________________ or a ________________ .
9. There is no movement when two forces are ________________ and

   ________________ .
10. Forces working together make a force ________________ .

REACHING OUT

Often work problems are given in sentences, not in arithmetic charts. To figure out the answers you first have to decide what you know from the sentence (the force, the distance, or the work). Then you must do the arithmetic to find the part you do not know.

1. A person lifts a rock 2 meters off the ground with a force of 20 Newtons. How much work did the person do? ________________ Joules
2. You use a 20 Newton force to move an object 5 meters. How much work do you do?

   ________________ Joules
3. A person does 10 Joules of work to lift an object 1 meter. How much force did the person use?

   ________________ Newtons

What is friction?

23

friction [FRIK-shun]: force that opposes the motion of an object
lubricate [LOO-bruh-kayt]: to apply a material to reduce friction

LESSON 23 | What is friction?

It is easy to slide on ice. Ice is smooth.

It is harder to slide on a sidewalk. A sidewalk is rough.

Everyone knows that it is easier to slide on a smooth surface than on a rough surface. But did you ever ask why?

Here is an explanation:

All surfaces have bumps and scratches. Even "smooth" surfaces have them. Smooth surfaces just have much smaller bumps and scratches than rough surfaces do. Sometimes they are so tiny you need a microscope to see them. Do you need a microscope to see a sidewalk's bumps and scratches?

When one surface moves on another surface, the bumps and scratches rub together. The rubbing is called **friction** [FRIK-shun]. Friction makes movement more difficult.

Friction resists movement.

There is always some friction when surfaces move against one another. The amount of friction can be large or small. The rougher the surface, the greater the friction.

Friction not only builds resistance, but also builds heat. Heat is not always wanted. That is why we try to reduce friction.

There are two main ways to reduce friction:

- Rolling things instead of sliding them. The amount of friction used in rolling something is less than the amount of friction used to slide it.
- Lubricating [LOO-bruh-kayt-ing] the surfaces of things. When you **lubricate** something, you put something slippery on it to make it slide more easily. Oil and grease are the most often used lubricants.

Figure A

Bumps and scratches on surfaces rub against each other and cause friction. All surfaces have bumps and scratches.

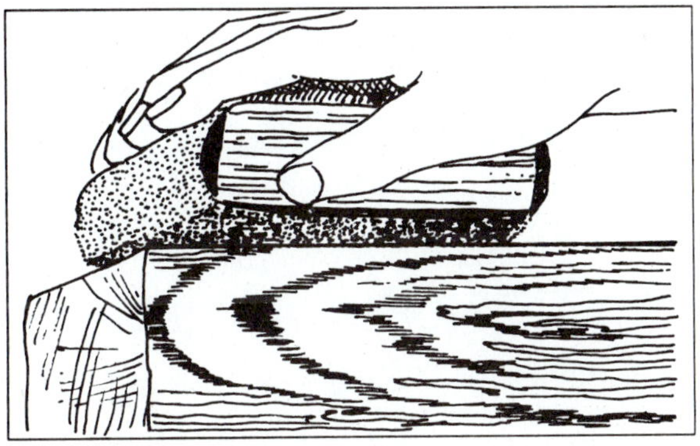

Figure B

Some surfaces have large bumps and scratches. You can see them easily.

Figure C

Some surfaces—such as metal—seem very smooth.

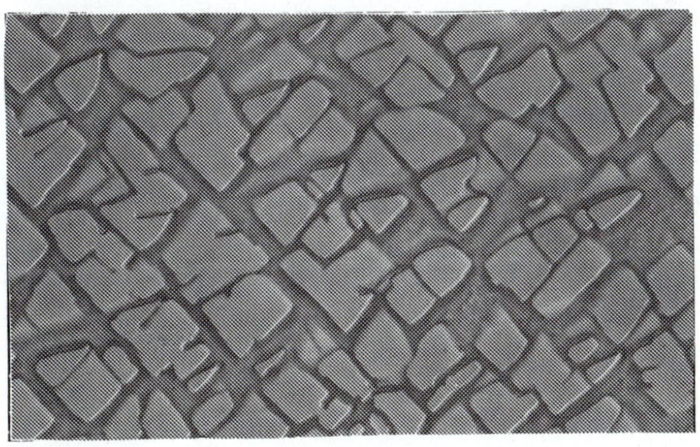

Figure D

However when examined closely you can see that they have tiny bumps and scratches.

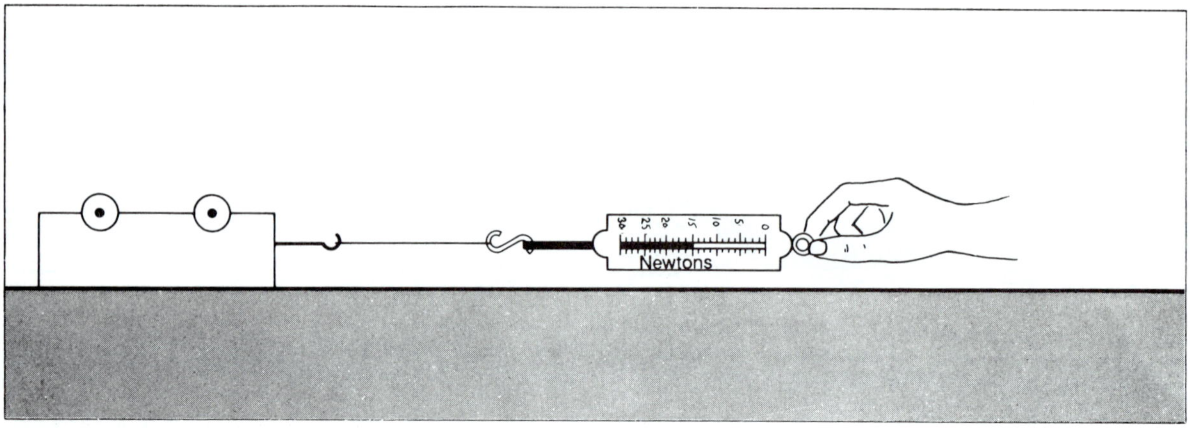

Figure E

1. How much force is pulling the laboratory carriage in Figure E? ____________

2. Are the wheels in use? ____________

3. This shows ____________ friction.
   sliding, rolling

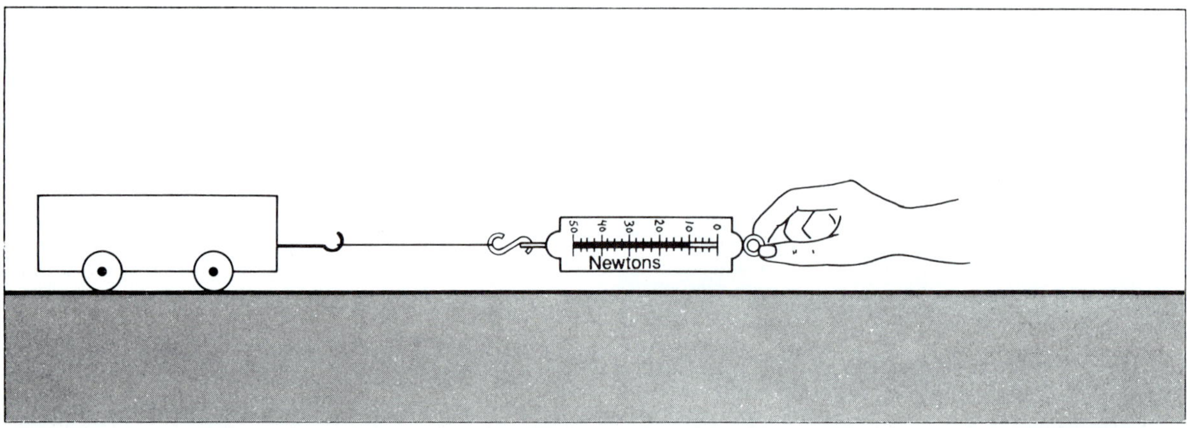

Figure F

4. a) Now, how much force is being used in Figure F? ____________

   b) This is ____________ force than was needed in Figure E.
   more, less

   c) Why is less force needed in Figure F? ____________

   ____

5. This is an example of ____________ friction.
   sliding, rolling

6. Rolling friction is ______________ than sliding friction.
greater, less

7. Sliding friction is ______________ than rolling friction.
greater, less

8. With sliding, there is ______________ resistance than with rolling.
more, less

9. With rolling, there is ______________ resistance than with sliding.
more, less

10. With sliding, you do ______________ work.
more, less

11. With rolling, you do ______________ work.
more, less

12. With sliding, you do more work because you use ______________ force.
more, less

13. With rolling, you do less work because you use ______________ force.
more, less

14. Which builds more heat, sliding or rolling? ______________

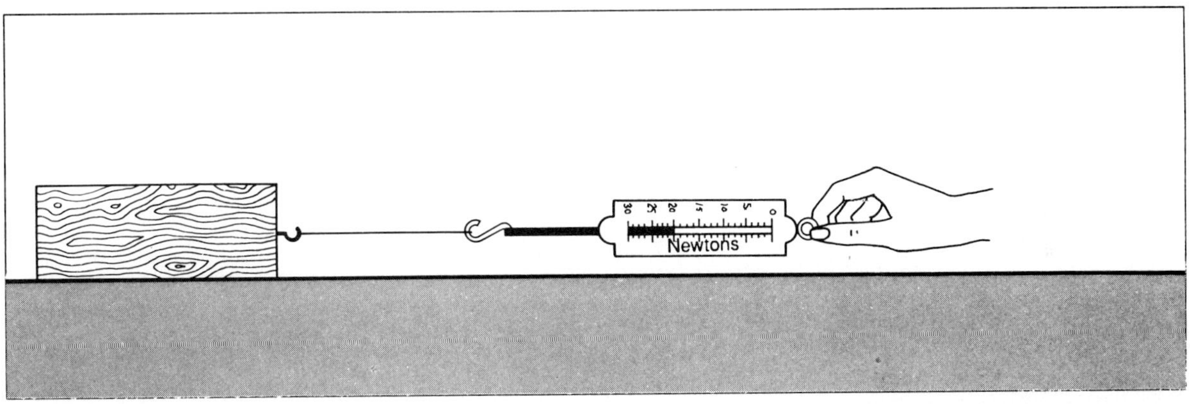

Figure G

15. How much force is pulling the piece of wood in Figure G? ______________

16. Is the surface lubricated? ______________

17. Now, how much force is needed to pull the wood in Figure H? (Turn the page.)

____

18. This is ______________ force than was needed in Figure G.
more, less

19. Why is less force needed? ______________

____

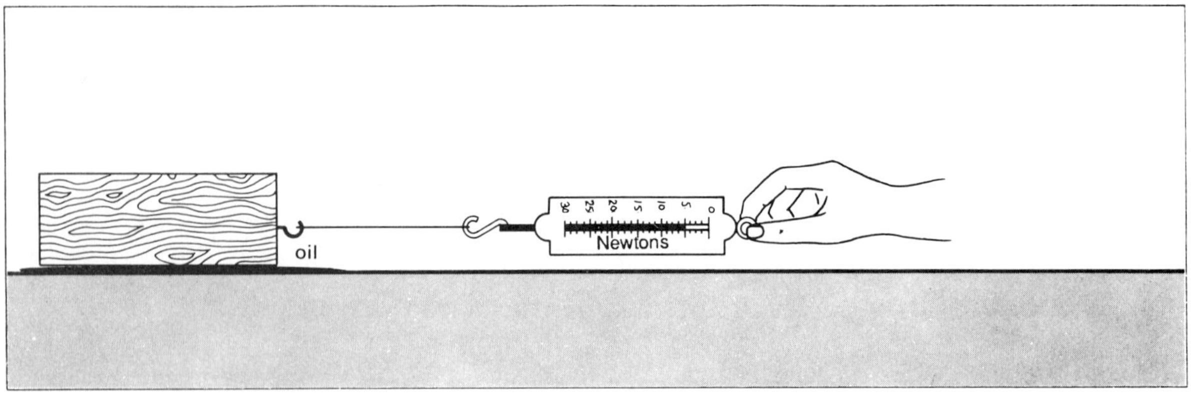

Figure H

20. Lubrication makes a surface ________________ .
(rougher, smoother)

21. Lubrication ________________ resistance.
(increases, decreases)

22. Lubrication ________________ friction.
(increases, decreases)

23. With lubrication you need ________________ force to move an object.
(more, less)

24. With lubrication, you do ________________ work.
(more, less)

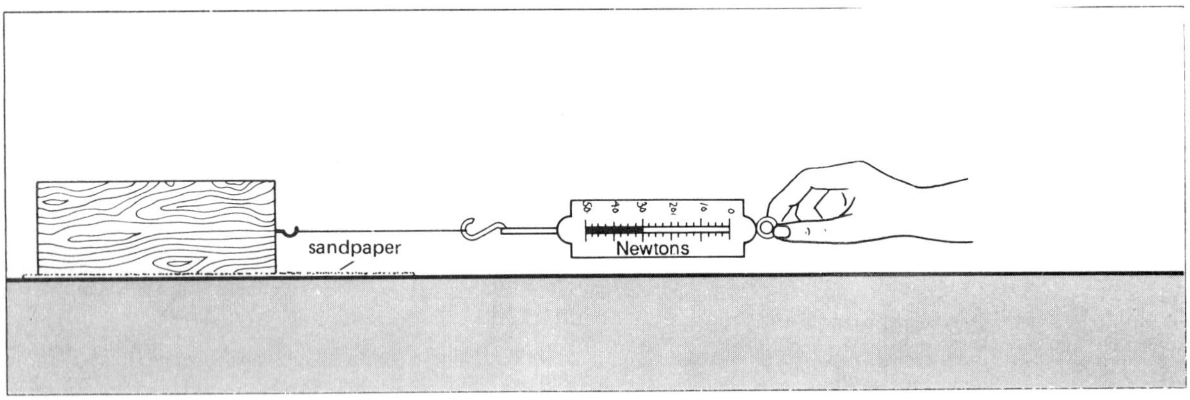

Figure I

25. How much force is needed to pull the wood in Figure I? ________________

26. This is ________________ force than was needed in Figure G.
(more, less)

27. Why is more force needed? ____

____

28. Sandpaper is a ________________ surface.
(rough, smooth)

29. Sandpaper ____________ friction.
increases, decreases

30. On sandpaper, you need ____________ force to move an object.
more, less

31. On sandpaper, you do ____________ work.
more, less

32. Which figure shows heat building up? ____________
G only, H only, I only, all three

33. Which figure shows the most heat building up? ____________
G, H, I

ADVANTAGES OF FRICTION

Figure J

If there were no friction a car's wheels would not move the car.

Figure K

Friction keeps a car from sliding down a hill.

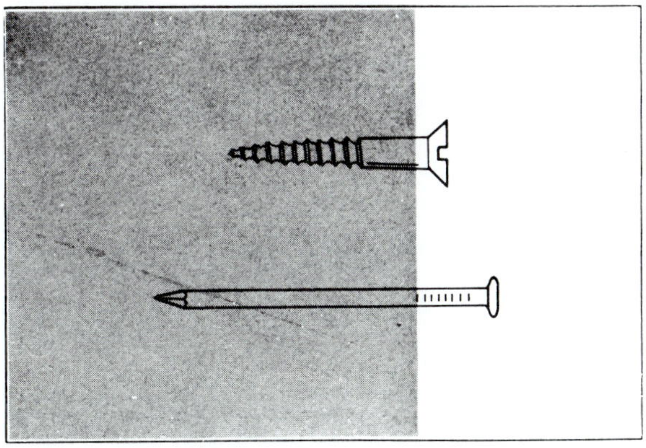

Figure L

The force of friction keeps nails and screws in wood.

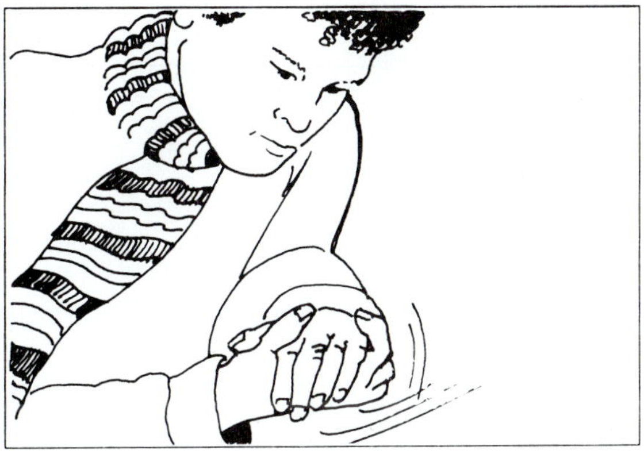

Figure M

The heat from friction can heat things up.

TRUE OR FALSE

In the space provided, write "true" if the sentence is true. Write "false" if the sentence is false.

_________ **1.** Friction always builds heat.

_________ **2.** Smooth surfaces build more heat than rough surfaces do.

_________ **3.** Oil and grease reduce friction.

_________ **4.** Sand is a lubricant.

_________ **5.** More friction means more work. [Remember: $W = F \times D$]

_________ **6.** Sandpaper is smoother than glass.

_________ **7.** Rolling friction is greater than sliding friction.

_________ **8.** It takes more work to move something on a lubricated surface.

_________ **9.** All surfaces are smooth.

_________ **10.** You would oil a bicycle wheel to increase its friction.

REACHING OUT

In regions where it snows, ice and snow on the roads can be very dangerous. Cars slide out of control on the ice. What can be done to make it safer to drive?

____

____

____

____

What is density?

24

density: the mass of a given volume
volume: a measure of the amount of space matter takes up

LESSON 24 What is density?

Rocks are heavy. A small box of rocks has a large mass. Yet a box of the same size containing corks has a small mass.

How can objects that are the same size have different masses? Scientists explain it this way:

All matter is made up of atoms. All atoms have mass. Some matter has more massive atoms than others. The more massive the atoms, the greater the mass.

Matter has mass. It also takes up space. The amount of space that matter takes up is called volume.

The atoms of some kinds of matter are more tightly packed than the atoms of other kinds. More atoms in a volume mean more mass in that volume.

Two properties of matter are important in the pull of gravity:

- the mass of each atom
- the number of atoms in a given volume

The mass of a given volume of matter is called its **density**.

It is the <u>differences</u> in densities that give objects of the same size different masses.

For example, gold is about twice as dense as copper. This means that if a piece of copper has a mass of 50 grams, then a piece of gold the same size would have a mass of 100 grams.

How do you figure density? It is easy. Use this simple formula.

$$\text{Density} = \frac{\text{Mass}}{\text{Volume}}$$

FIGURING VOLUME

Volume is measured in cubic units. One common cubic unit is a cubic centimeter (cc).

The volume of an object shaped like a box can be found by multiplying three numbers:

length × width × height.

Example: What is the volume of a box 5 centimeters long, 3 centimeters wide, and 2 centimeters high?

volume = l × w × h
volume = 5 cm × 3 cm × 2 cm
(5 × 3 × 2 = 30)
volume = 30 cubic centimeters

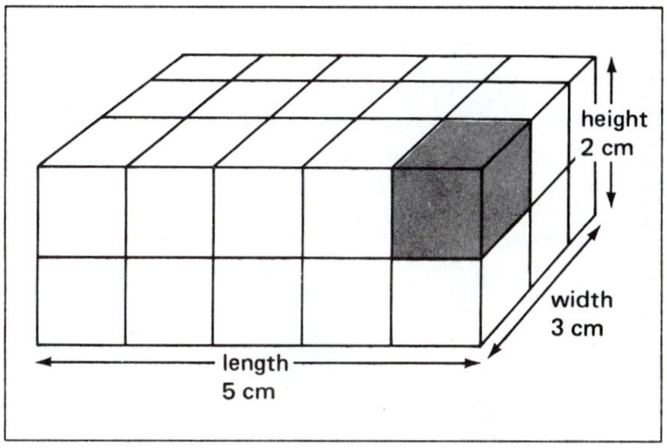

Figure A

This means that 30 one-cubic centimeter boxes can fit inside this box.

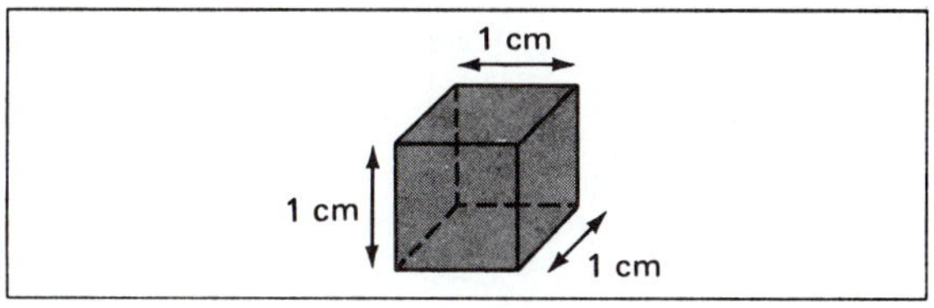

Figure B

FIGURING DENSITY

Now let us say that this same 30 cubic centimeter box has a mass of 120 grams. Let us find its density.

Density = $\frac{\text{mass}}{\text{volume}}$

Density = $\frac{\text{120 grams (g)}}{\text{30 cubic centimeters (cc)}}$
(120 ÷ 30 = 4)

Density = 4 grams per cubic centimeter
(4 g/cc)

Another way of saying this is:

Each cubic centimeter of this substance has a mass of 4 grams.

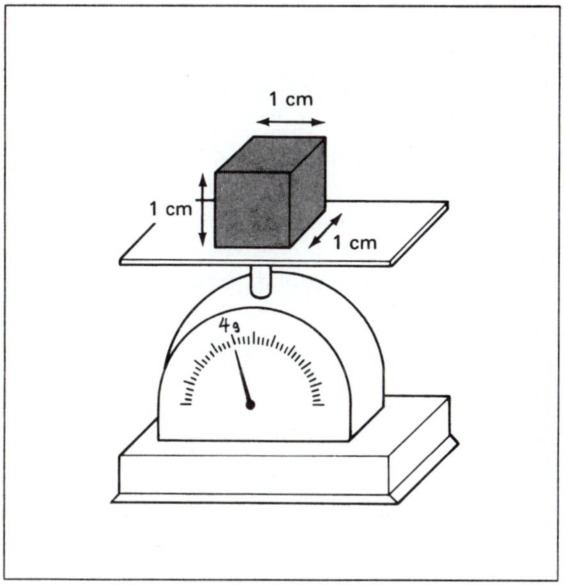

Figure C

COMPARING DENSITIES

Equal volumes of different substances have different masses.

Each one of these cubes has a volume of one cubic centimeter.

Each substance has a different mass.

Each substance has a different density.

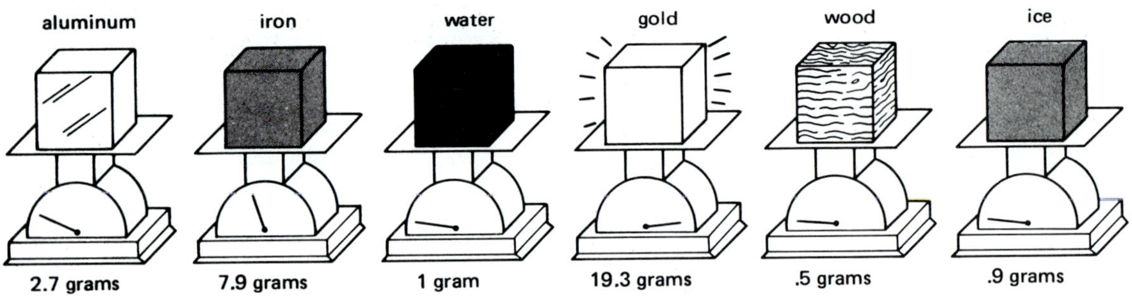

Figure D

Answer these questions.

1. Which substance has the greatest density? ________________

2. Which substance has the lowest density? ________________

3. Iron, aluminum, and gold are metals. Which of the metals has the greatest density?

   ____

4. Which of the metals has the lowest density? ________________

5. Anything that is <u>less</u> dense than water will float. Which two substances will float on water?

   ________________ ________________

UNDERSTANDING DENSITY

Figure E shows a solid iron cube and a solid aluminum cube. Study the figure. Then fill in the blanks.

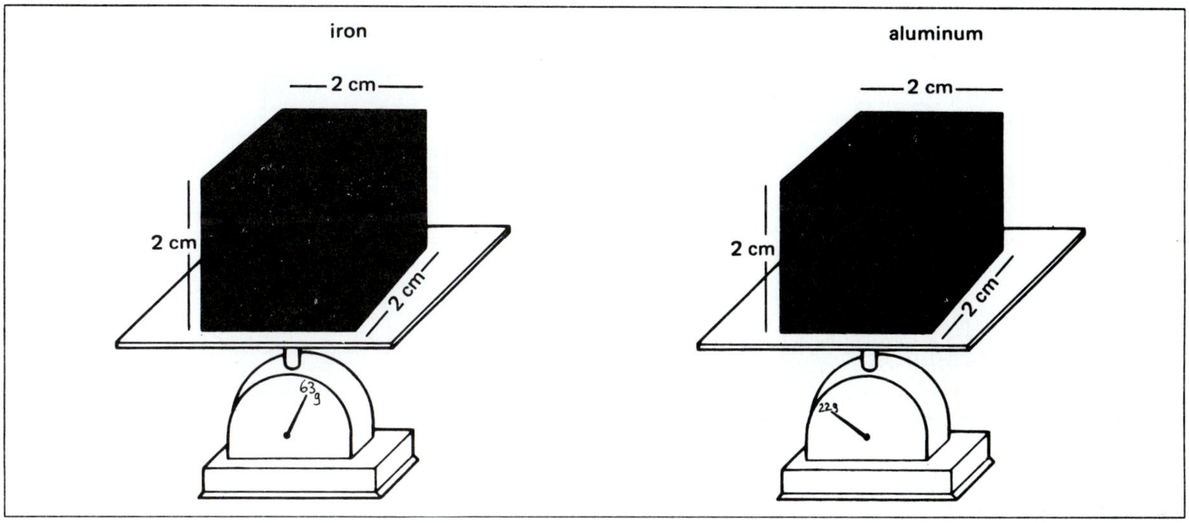

Figure E

1. The volume of the piece of iron is ______________ cubic centimeters.

2. The volume of the piece of aluminum is ______________ cubic centimeters.

3. The volumes ______________ (are, are not) the same.

4. They ______________ (do, do not) have the same mass.

5. The ______________ (aluminum, iron) has a greater mass.

6. Iron is ______________ (more, less) dense than aluminum.

7. Aluminum is ______________ (more, less) dense than iron.

8. Iron is ______________ (sometimes, always) more dense than aluminum.

9. Aluminum is ______________ (sometimes, always) less dense than iron.

10. Things made of iron ______________ (always, sometimes) are more massive than things made of aluminum.

11. Things made of aluminum ______________ (can, can never) have a greater mass than things made of iron.

} These are tricky. Think before answering!

UNDERSTANDING VOLUME AND DENSITY

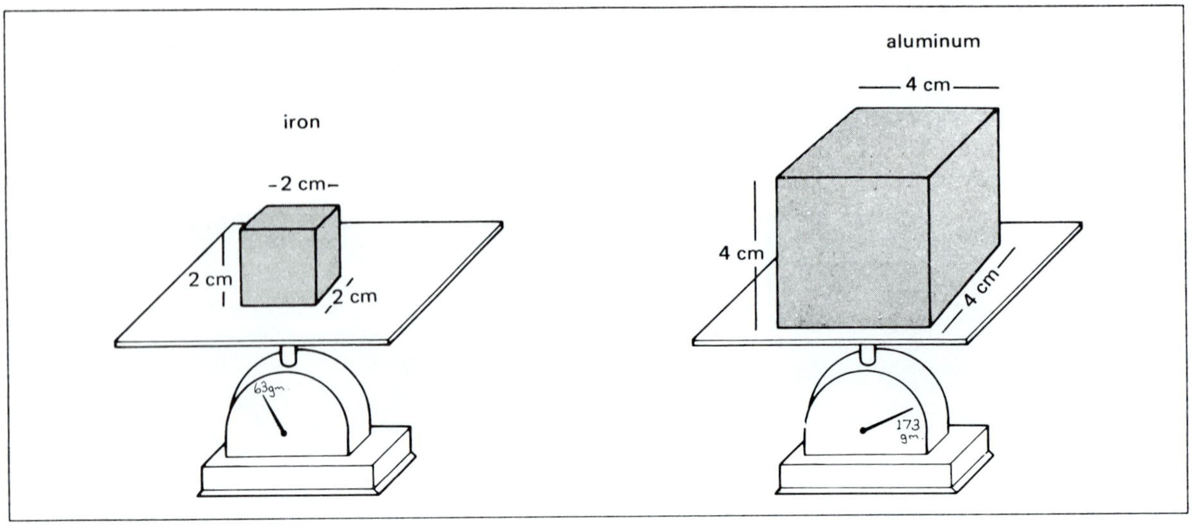

Figure F

1. We know that iron is ______________ (more, less) dense than aluminum.

2. A volume of iron has ______________ (larger, smaller) mass than an equal volume of aluminum.

Now look at Figure F.

3. In Figure F the aluminum has a ______________ (larger, smaller) mass than the iron.

4. This ______________ (does, does not) mean that aluminum is more dense than iron.

5. The aluminum is more massive because there is a ______________ (larger, smaller) volume of aluminum than there is of iron.

MATCHING

Match each term in Column A with its description in Column B. Write the correct letter in the space provided.

	Column A	Column B
______	**1.** mass	**a)** formula for density
______	**2.** volume	**b)** the amount of matter in a substance
______	**3.** density	**c)** formula for volume
______	**4.** length × width × height	**d)** the mass of a given volume
______	**5.** $\frac{\text{mass}}{\text{volume}}$	**e)** the amount of space matter takes up

FINDING DENSITIES

For each of the examples below, first, find the volume. Then find the density. The first one has been done for you.

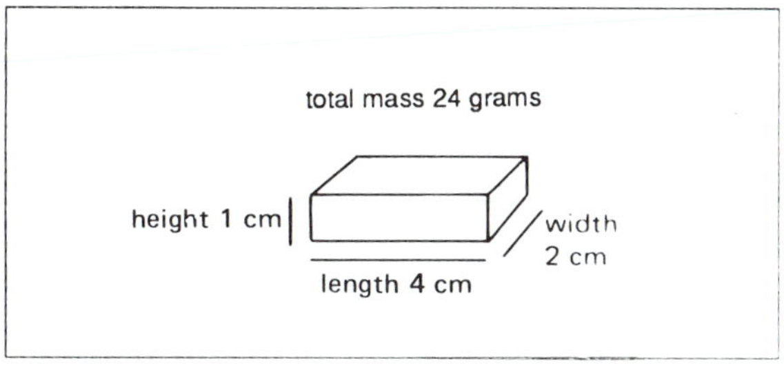

Figure G

1. Finding volume

Volume = l × w × h

Volume = 4 × 2 × 1

Volume = 8 cubic centimeters (cc)

Finding density

Density = $\frac{\text{mass}}{\text{volume}}$

Density = $\frac{\text{24 grams}}{\text{8 cubic centimeters}}$

Density = 3 grams/cubic centimeter

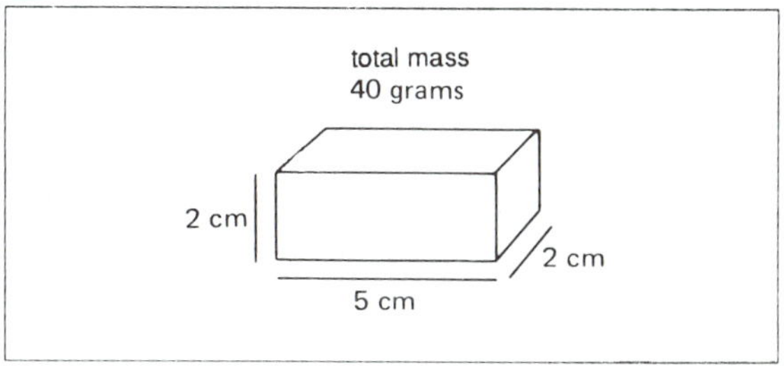

Figure H

2. Finding volume

Volume = l × w × h

Volume = ______________

Volume = ______________

Finding density

Density = $\frac{\text{mass}}{\text{volume}}$

Density = $\frac{\text{\_\_\_\_\_\_\_\_ g}}{\text{\_\_\_\_\_\_\_\_ cc}}$

Density = ______________ grams/cc

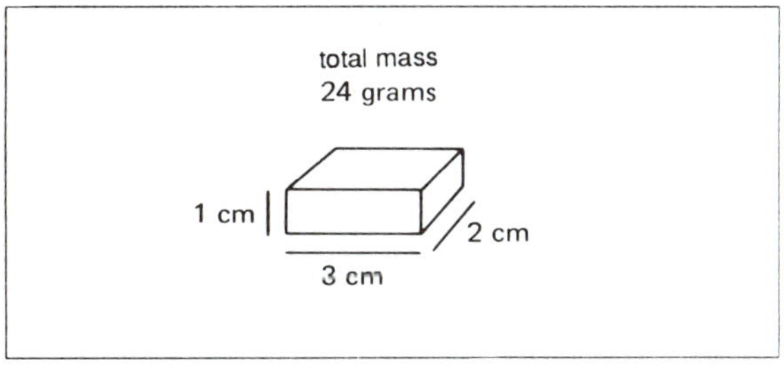

Figure I

3. Volume = l × w × h

Volume = ______________

Volume = ______________ cc

Density = $\frac{\text{mass}}{\text{volume}}$

Density = $\frac{\text{\_\_\_\_\_\_\_\_ g}}{\text{\_\_\_\_\_\_\_\_ cc}}$

Density = ______________ g/cc

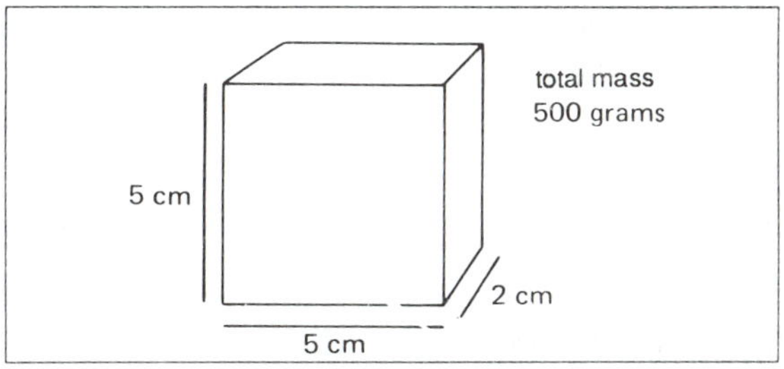

Figure J

4. Volume = l × w × h

Volume = ______________

Volume = ______________ cc

Density = $\frac{\text{mass}}{\text{volume}}$

Density $\frac{\text{\_\_\_\_\_\_\_\_ g}}{\text{\_\_\_\_\_\_\_\_ cc}}$

Density = ______________ g/cc

TRUE OR FALSE

In the space provided, write "true" if the sentence is true. Write "false" if the sentence is false.

__________ **1.** All matter has mass and takes up space.

__________ **2.** All atoms have the same mass.

__________ **3.** An atom of iron is more massive than an atom of aluminum.

__________ **4.** A pile of feathers can be more massive than a pile of rocks.

__________ **5.** Density has to do only with mass.

__________ **6.** Volume tells us how massive matter is.

__________ **7.** Ice floats on water.

__________ **8.** Ice is more dense than water.

__________ **9.** Equal volumes always mean equal masses.

__________ **10.** A centimeter is a measure of volume.

REACHING OUT

A measuring cup or graduated cylinder measures liquid volume.

One measurement of volume is the cubic centimeter. Another is the milliliter.
One milliliter (ml) = one cubic centimeter.

1. What is the total volume of the rock and liquid in this graduated cylinder? _______________

2. How can you use a graduated cylinder filled with water to find the volume of an odd-shaped object like a rock?

____

____

____

____

____

____

____

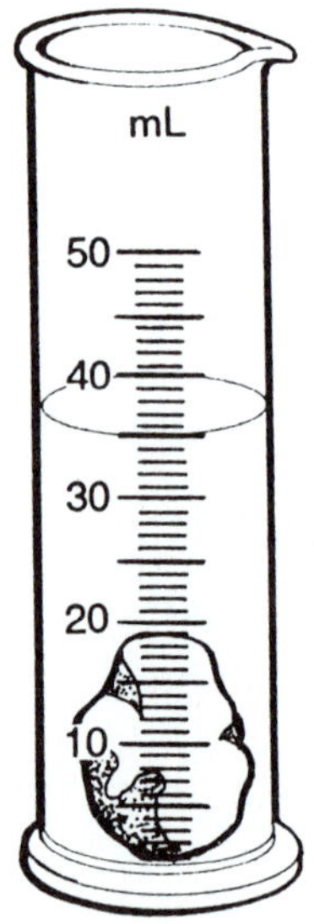

Figure K

What is pressure?

25

atmosphere [AT-mus-fir]: layer of gases that surrounds the earth
manometer [muh-NAHM-uh-tur]: an instrument that measures water pressure
pressure: a force on a definite area

LESSON 25 What is pressure?

The date was April 10, 1963. The American submarine *Thresher* was making test runs off the coast of Boston. There were 129 sailors aboard.

Radio contact gave no hint of trouble. Suddenly, there was silence. The *Thresher* was not heard from again.

Planes and ships searched the area, but no trace of the sub was found.

Finally, an underwater camera discovered the terrible truth. Pictures showed pieces of the *Thresher* spread out on the ocean floor. The submarine had gone too deep. The water pressure had crushed its hull. Everyone aboard had died.

What is **pressure**?

PRESSURE IS A FORCE ON A DEFINITE AREA.

Pressure depends on two things:

- The amount of force on an object.
- The size of the area the force pushes or pulls on.

Pressure is measured in units of force on a given area. For example, a force of 10 Newtons pushing on one square centimeter has a pressure of 10 Newtons per square centimeter. We can also write this as 10 N/sq cm.

Pressure can be caused by the weight of matter. The more matter there is pressing on an object, the greater the pressure. Usually, the denser the matter is, the greater the pressure.

Take water, for example. Water is more dense than air. Therefore, water pressure is greater than air pressure. The deeper we go in water, the more water there is above us. The mass of water above us gets greater and greater. Therefore, the pressure becomes greater and greater.

Natural pressure is caused by gravity pulling upon matter. Solids press only downward. Liquids and gases are different. Liquids and gases press in all directions. At a given point in a liquid or a gas, the pressure is the same in every direction.

TRY THIS!

Outstretch your hand, palm up. Pile several books on your hand—one at a time. Notice how the pressure changes.

Figure A

1. A book ______________ matter.
   is, is not

2. A book ______________ have weight.
   does, does not

3. Weight ______________ cause pressure.
   does, does not

4. You feel the least pressure with ______________ book(s).
   one, many

5. As you add more books, you feel ______________ pressure.
   more, less

6. You feel more pressure because there is ______________ mass.
   more, less

7. The area of your palm ______________ changed.
   has, has not

Now, let's change the area.

Pile two of the books on your palm. Feel the pressure.

Now, balance the two books on three fingertips only.

Figure B

8. The weight of the books ______________ changed.
   has, has not

9. The weight of the books is now spread over a ______________ area.
   larger, smaller

10. The pressure on each of your fingertips is ______________ than it was on your palm.
    greater, less

What conclusions can we make about pressure?

11. If we keep the weight (force) the <u>same</u> and spread it over a <u>smaller</u> area we

_____________________ the pressure.
increase, decrease

12. If we keep the weight (force) the <u>same</u> and spread it over a <u>larger</u> area we

_____________________ the pressure.
increase, decrease

13. If we keep the area the <u>same</u> and <u>increase</u> the force we _____________________ the pressure.
increase, decrease

14. If we keep the area the <u>same</u> and <u>decrease</u> the force we _____________________ the pressure.
increase, decrease

15. Solids, like the books, cause a(n) _____________________ pressure.
upward, downward

MEASURING WATER PRESSURE

The instrument in Figure C is called a manometer [muh-NAHM-uh-tur].

A **manometer** measure liquid pressure. The liquid in the "U" tube rises as the pressure becomes greater.

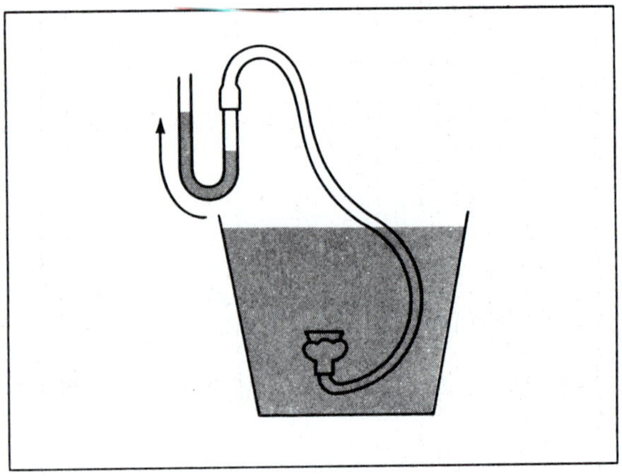

Figure C

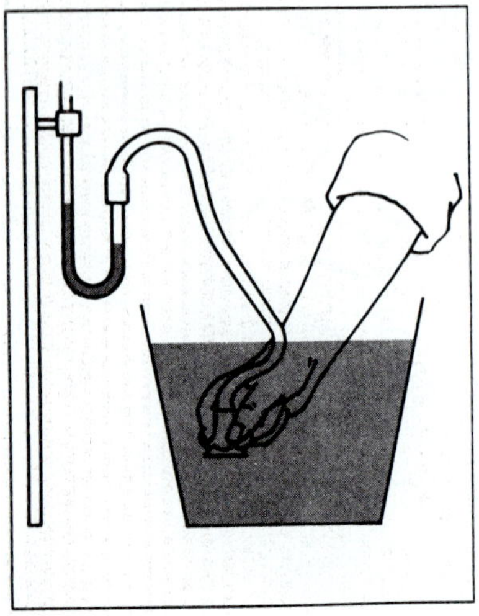

Figure D

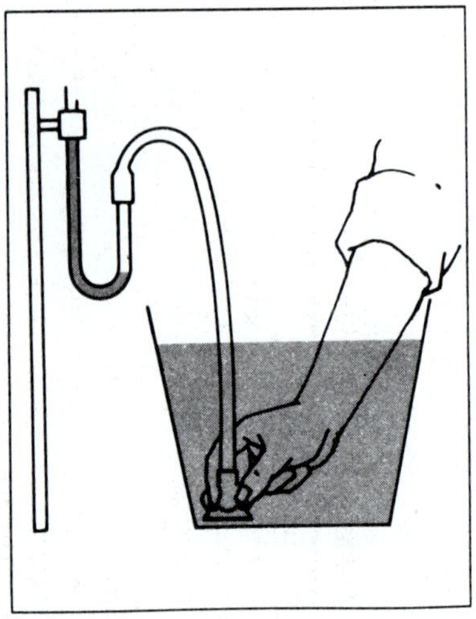

Figure E

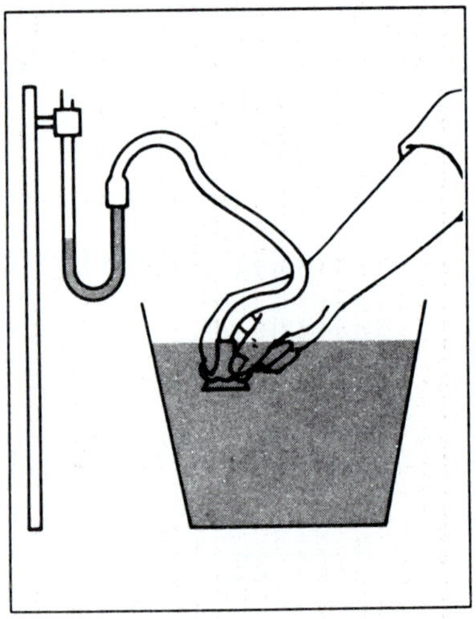

Figure F

Figure D, E, and F show a manometer in water. In each one, the manometer "bulb" is at a different <u>depth</u>.

Study each figure. Then fill in the blanks.

1. The water level on the left side of the "U" tube is **a)** lowest in ________________ .

   b) highest in ________________ .

2. Water pressure is **a)** greatest in ________________ .

   b) least in ________________ .

3. Pressure is greatest at the ________________ of the water.
   top, bottom

4. Pressure is lowest at the ________________ of the water.
   top, bottom

5. Water pressure ________________ the deeper you go in water.
   increases, decreases

Figures G, H, and I show a manometer in water. In each one, the manometer bulb is in a different <u>position</u>. Study each figure. Then fill in the blanks.

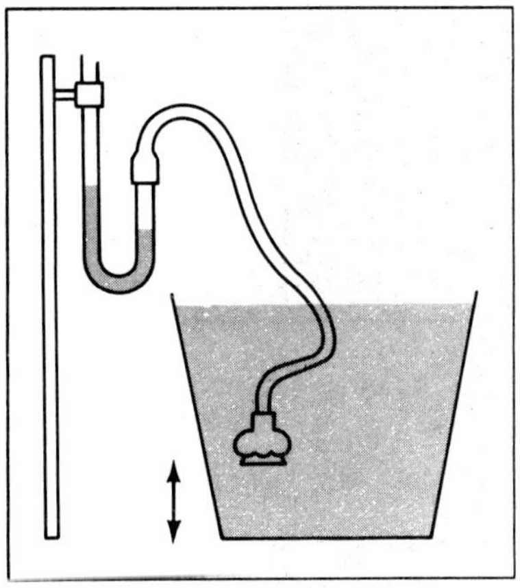

Figure G

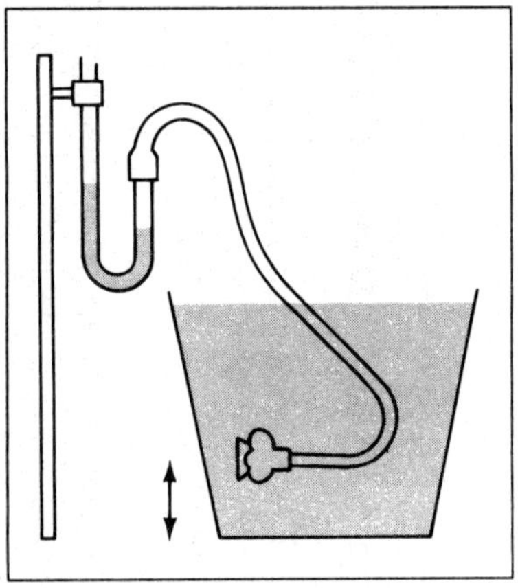

Figure H

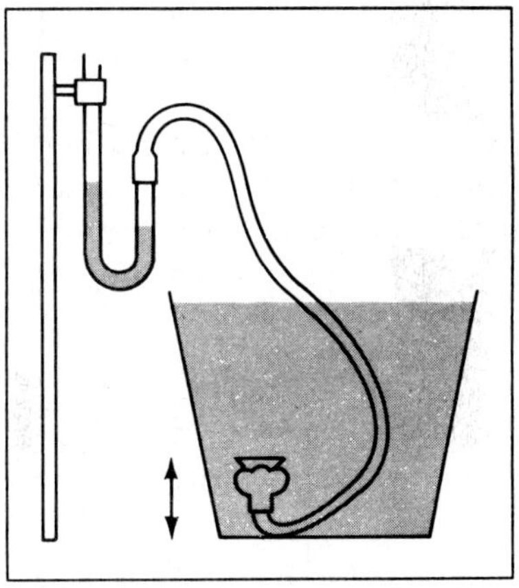

Figure I

6. In Figures G, H, and I the bulb of the manometer is at ________________ .
   one depth, different depths

7. Changing the position of the bulb ________________ changed the water level in the tube.
   has, has not

8. This means that the pressure ________________ changed.
   has, has not

9. This demonstration shows that water presses ________________ .
   only downward, only upward, on all sides

10. At a given point, water pressure ________________ the same in every direction.
    is, is not

ABOUT AIR PRESSURE

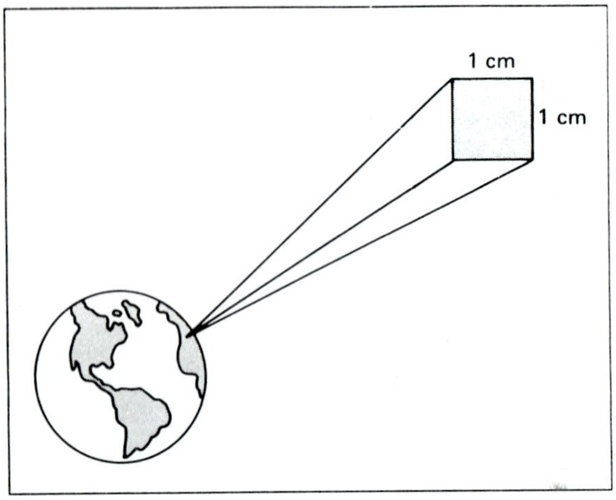

Figure J

The **atmosphere** [AT-mus-fir] is the layer of gases that surrounds the earth. It is about 1000 kilometers (600 miles) thick.

Gases do not weigh much. However, a layer of gases this thick is quite heavy.

At sea level, air presses down with a force of about 10 Newtons on every square centimeter. (10 N/sq cm)

Does that sound like a lot? Look at just one sheet of paper in this book. The total force on this sheet of paper is over 5000 N, (or 1130 lbs).

MEASURING AREAS AND FIGURING TOTAL FORCES

Measure the rectangles in Figures K, L, M, and N with your metric ruler. Find the area of each. (Area = length × width) Then figure the total force that is pressing down on each area.

Remember, air pressure is 10 Newtons per square centimeter (10 N/sq cm).

Example 1 has been worked out for you.

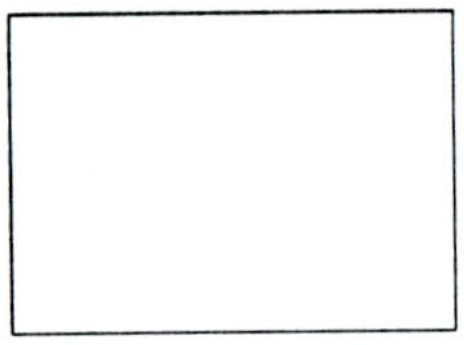

Figure K

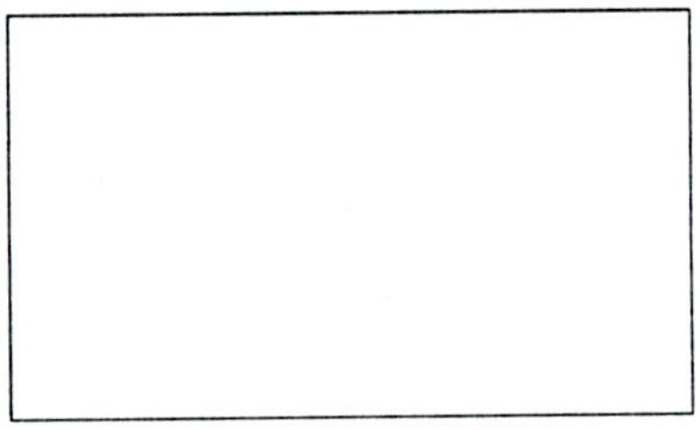

Figure L

1. length = 3 cm
width = 2 cm
area (l × w) = 6 sq cm

Total force = area × pressure

Force = 6 sq cm × 10 N/sq cm

Force = 60 N

2. length = ______________ cm

width = ______________ cm

area = ______________ sq cm

Total force = ______________ N

What would be the total force if the pressure were 20 N/sq cm?

____

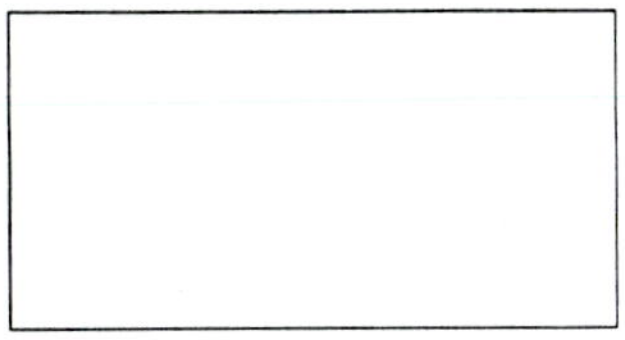

Figure M

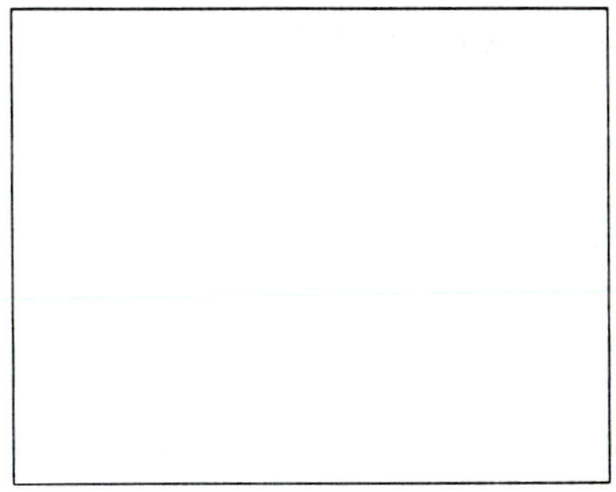

Figure N

3. length = ____________________ cm

width = ____________________ cm

area = ____________________ sq cm

Total force = ____________________ N

What would be the total force if the pressure were 30 N/sq cm? __________

4. length = ____________________ cm

width = ____________________ cm

area = ____________________ sq cm

Total force = ____________________ N

What would be the total force if the pressure were 5 N/sq cm? __________

FILL IN THE BLANK

Complete each statement using a term or terms from the list below. Write your answers in the spaces provided. Some answers may be used more than once.

in all directions	downward	pressure	more
higher	deeper	weight	heavier
area	gravity	force	

1. A force on a certain area is called ____________________ .
2. Pressure comes from the ____________________ of matter.
3. The force that pulls matter downward is called ____________________ .
4. Solids press only ____________________ .
5. Liquids and gases press ____________________ .
6. Pressure increases when there is ____________________ matter or ____________________ matter.
7. The deeper you go in water, the ____________________ water there is above you.
8. Water pressure becomes greater the ____________________ you go.
9. Gravity becomes weaker the ____________________ you go in the atmosphere.
10. Pressure is measured in units of ____________________ over a given ____________________ .

MATCHING

Match each term in Column A with its description in Column B. Write the correct letter in the space provided.

	Column A	Column B
______	**1.** gravity	**a)** press in all directions
______	**2.** solids	**b)** a pressure measurement
______	**3.** liquids and gases	**c)** press only downward
______	**4.** square centimeter	**d)** an area measurement
______	**5.** 1 gram/square centimeter	**e)** pulls things downward

REACHING OUT

Where is air pressure greater, high in the atmosphere or on the ground? ______________

____

Why? ____

____

____

____

Figure O

Why do objects float?

26

ballast [BAL-ust] **tanks:** tanks on a submarine that make the submarine raise or lower its level in water

LESSON 26 | Why do objects float?

When you throw a rock into water, it sinks. It sinks because a rock is more dense than water. When you throw wood into water, it floats. It floats because wood is less dense than water.

Objects that are more dense than water sink in water.

Objects that are less dense than water, float on water.

Here is another comparison.

The density of water is 1 gram per cubic centimeter.

The density of steel is nearly 8 grams per cubic centimeter. This is eight times more dense than water.

Steel, then, should sink in water—and it does. Yet a steel ship that weighs 100 thousand tons or more floats on water.

How can this be explained? Why do steel ships float?

The answer is simple. A ship may be made of steel. But only a small part of its volume is steel. Because of the shape of the hull, most of the ship's volume is air. Air is less dense than water.

The density of a ship, then, is not just the density of the metal. It is the density of the metal, the air in the hull, and all the other things in the ship. Altogether, the density is less than the density of water. Objects that are less dense than water float in water. That is why a ship floats.

UNDERSTANDING FLOATING

A flat piece of metal has no air trapped within its border. All its volume is metal. All its density is metal.

Figure A

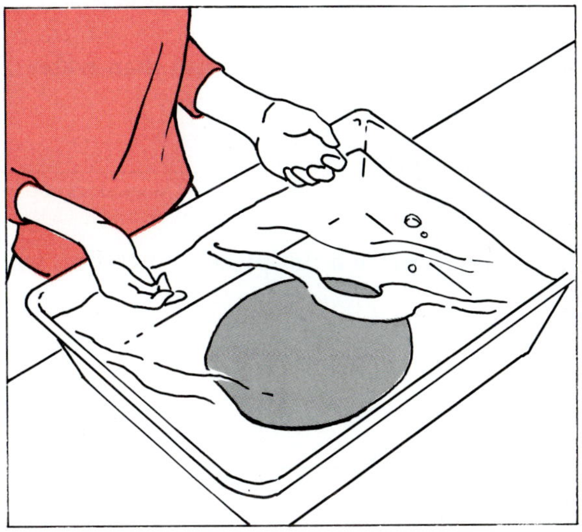

Figure B

Metal is more dense than water. It sinks in water.

What happens when you shape the metal like a bowl—or a ship?

Figure C

Figure D

The new shape has a volume of metal—and a great deal of air.

The new shape has a new density. It is the density of the steel as well as the air. Air has a low density.

Altogether, the density is less than the density of water. The shaped steel floats on water.

COMPLETING SENTENCES

Choose the correct word or term for each statement. Write your choice in the spaces provided.

1. Objects that are more dense than water ________________ in water.
   float, sink

2. Objects that are less dense than water ________________ in water.
   sink, float

3. A flat piece of metal ________________ have air trapped within its sides.
   does, does not

4. If you place a flat piece of metal on water, it will ________________ .
   sink, float

5. A ship is shaped like a ________________.
   table top, bowl

6. A bowl has ________________ volume than the flat metal it is made of.
   more, less

7. Most of the volume of a ship is ________________ .
   water, air, metal

8. Air is mostly ________________ .
   liquids, gases

9. Gases are ________________ dense than metal.
   more, less

10. Filling a ship with water would make it ________________ dense.
    more, less

MATCHING

Match each term in Column A with its description in Column B. Write the correct letter in the space provided.

	Column A	Column B
________	1. volume	a) measured in grams per cubic centimeter
________	2. density	b) measured in Newtons per square centimeter
________	3. pressure	c) length x width x height
________	4. floats	d) an object more dense than water
________	5. sinks	e) an object less dense than water

CHANGES IN DENSITY

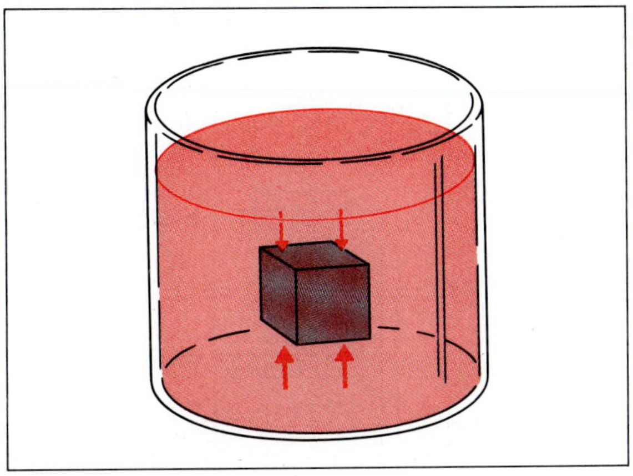

Figure E

The deeper we go in water, the more water there is above us. The mass of the water above us gets greater and greater. Therefore, the pressure becomes greater and greater. This pressure squeezes the water making it more dense. Some objects are too dense to float on water's surface. These objects sink down to deeper water. The objects will continue to sink until they reach water that is as dense as they are. There, the objects will float.

COMPLETING SENTENCES

Choose the correct word or term for each statement. Write your choice in the spaces provided.

1. The water at the surface of the container is ______________ (higher, lower) than the water at the bottom of the container.

2. The water at the bottom of the container is ______________ (higher, lower) than the water at the top of the container.

3. The water is pressing ______________ (upward, downward) on the top surface.

4. The water is pressing ______________ (upward, downward) on the bottom surface.

5. The pressure of the water at the top of the container is ______________ (greater, less) than the pressure on the bottom of the container.

6. The pressures ______________ (are, are not) equal.

7. The water is ______________ (more, less) dense at the top of the container.

8. The water is ______________ (more, less) dense at the bottom of the container.

9. The object ______________ (does,does not) float at the top of the container.

10. The object ______________ (does, does not) float at the bottom of the container.

DENSITY AND SEPARATING SUBSTANCES

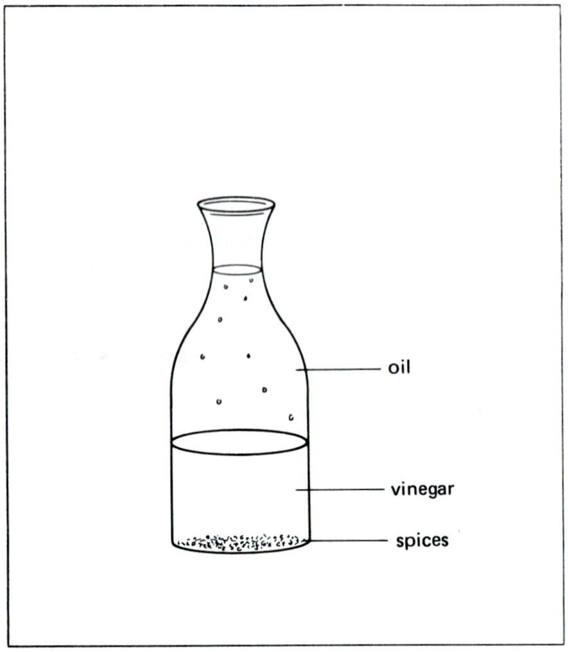

Figure F

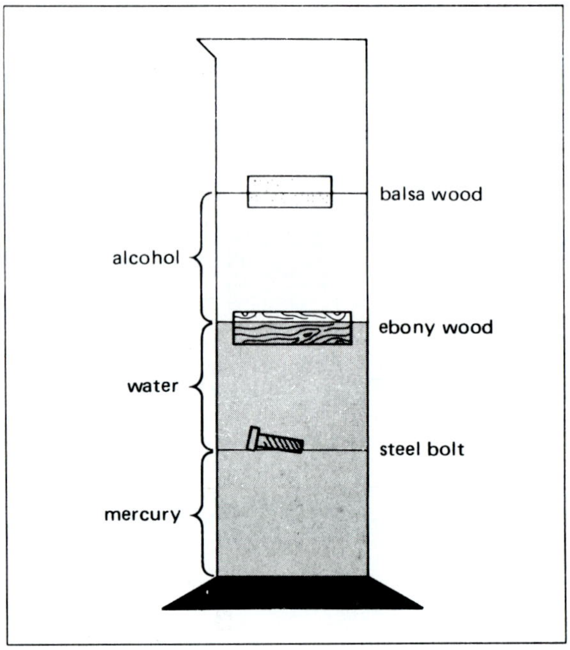

Figure G

Figure F shows salad dressing before it is mixed.

1. The parts are separated because they have different ______________ . (sizes, densities)

2. Which part is most dense? ______________ How do you know? ______________

3. Which part is the least dense? ______________ How do you know? ______________

Study the mixture in Figure G. Then answer the questions.

4. Name the solids in the glass. ______________

5. Name the liquids. ______________

6. Which solid is the most dense? ______________

7. Which solid is the least dense? ______________

8. Which liquid is the most dense? ______________

9. Which liquid is the least dense? ______________

10. Which of these substances have densities greater than water? ______________

11. Which have densities less than water? ______________

12. Which substance is the most dense? ______________

13. Which substance is the least dense? ______________

SUBMARINES

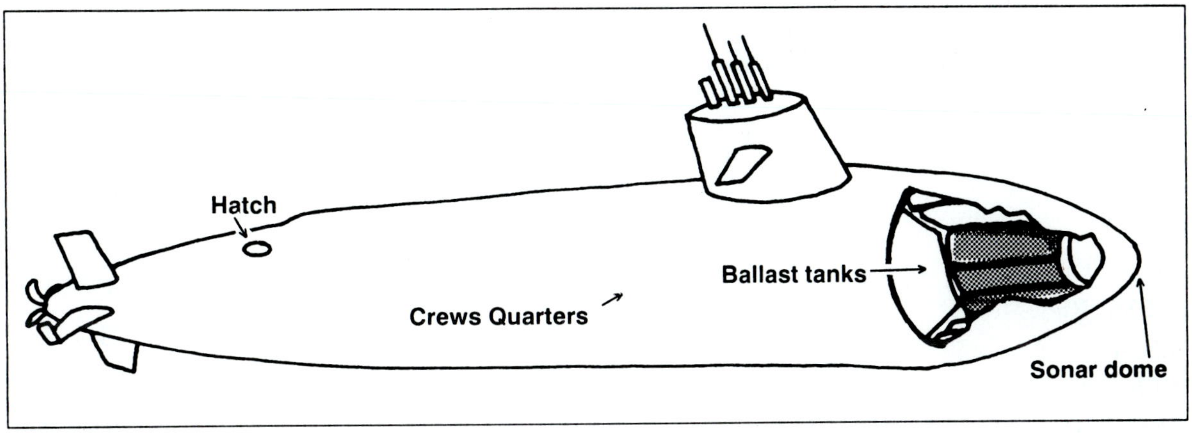

Figure H

A submarine has special tanks called **ballast** [BAL-ust] tanks. Ballast tanks are used to make the submarine heavier or lighter. When the ballast tanks are filled with seawater, the submarine submerges, or goes under water. When the water is forced out of the ballast tanks, the submarine rises and surfaces.

1. Adding ballast water makes a submarine ______________ dense.
   more, less
2. Forcing ballast water out makes a submarine ______________ dense.
   more, less
3. When a submarine is submerging its density is ______________ than water.
   greater, less
4. When a submarine is surfacing its density is ______________ than water.
   greater, less
5. When a submarine is submerging, its ballast tanks are ______________ .
   full, empty

TRUE OR FALSE

In the space provided, write "true" if the sentence is true. Write "false" if the sentence is false.

__________ 1. A bar of steel will float on water.

__________ 2. Then density of steel is greater than water.

__________ 3. A steel ship will float on water.

__________ 4. A steel ship is mostly metal.

__________ 5. The density of a steel ship is greater than the density of water.

__________ 6. Part of every floating object is below the water.

__________ 7. Water pressure increases the deeper you go.

__________ 8. An object will float in water if it is less dense than water.

__________ 9. An object will sink in water if it is less dense than water.

__________ 10. A submarine can change its density.

WORD SEARCH

The list on the left contains words that you have used in this Lesson. Find and circle each word where it appears in the box. The spellings may go in any direction: up, down, left, right, or diagonally.

FLOAT
BALANCE
PRESSURE
AREA
WEIGHT
DENSITY
VOLUME
MASS
MATTER
ENERGY
DISPLACE

G	D	O	N	Y	A	L	E	E	W	P
W	B	A	R	Y	G	R	E	N	E	L
I	A	V	O	L	U	M	E	R	I	E
E	L	F	L	O	A	T	U	S	G	E
L	A	M	A	S	T	S	E	A	H	C
E	N	M	A	S	S	R	E	N	T	A
S	C	E	A	E	P	R	O	Z	W	L
L	E	S	R	T	A	R	S	Y	S	P
I	P	P	A	Q	T	L	P	T	V	S
I	Y	T	I	S	N	E	D	U	R	I
E	L	N	E	P	S	O	R	A	R	D

REACHING OUT

Gases separate by density. So do liquids. Even a solid, when it is in a gas or a liquid, will find its level by density. Yet, solids of different densities do not separate by density. For example, aluminum is less dense than iron. Wood is even less dense than iron or aluminum. Yet, when these substances are placed one on top of another, they do not separate by densities.

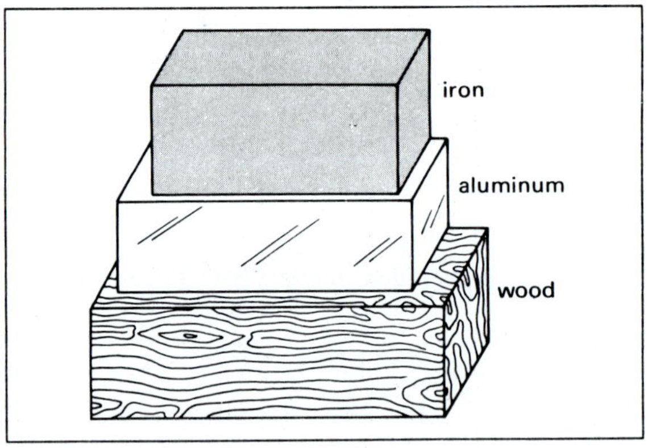

Figure I

Why? ____

____

____

____

What are simple machines?

27

complex machines: complicated machines made up of more than one simple machine
machine: any devise that can change strength of a force , its speed, or its direction
simple machines: basic machines; inclined plane, wedge, screw, lever, wheel and axle, and pulley

LESSON 27 | What are simple machines?

Muscles do many jobs. All day long they push and pull for you. However, many jobs need more force than muscles alone can give. For example, you can lift a spare tire out of an car trunk, but you need a jack to lift the whole car.

A **machine** can make a force stronger. It can change a small force into a large force. This means that a small force can move heavy objects.

A machine can be big and complicated, or it can be simple. The most basic machines are known as simple machines.

There are six kinds of **simple machines**. They are:

- the inclined plane
- the wedge
- the screw
- the lever
- the wheel and axle
- the pulley

Most of us do not usually think of these as machines, but they are. Each one can make a force stronger. Anything that can make a force stronger is a machine.

The six simple machines can be combined in many ways to make complicated machines. Complicated machines are often called **complex machines**. If you examine a complex machine, you will see a combination of simple machines. A crane is an example of an complex machine.

Machines do not reduce the amount of work. Machines just divide work into several smaller and easier jobs. It makes no difference whether you use your muscles alone or get help from a machine. In the end, the amount of work you do is the same.

THE SIX SIMPLE MACHINES

Figure A *the inclined plane*

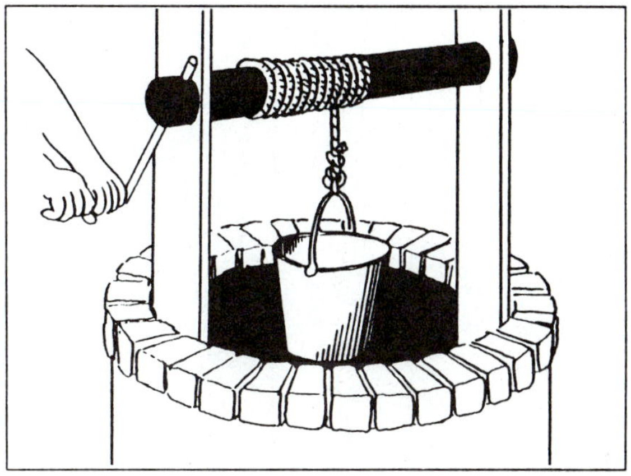

Figure B *the wheel and axle*

Figure C *the pulley*

Figure D *the lever*

Figure E *the wedge*

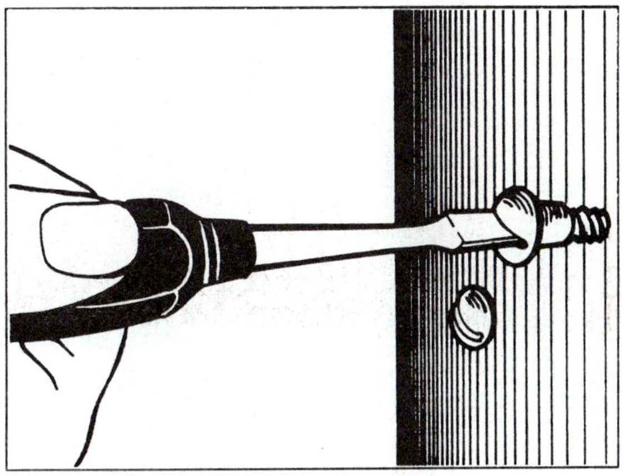

Figure F *the screw*

WHAT DO MACHINES DO?

Figure G *A machine can change direction.*

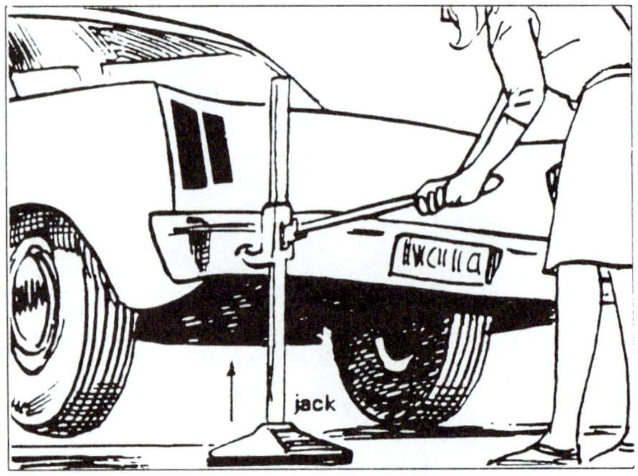

Figure H *A machine can multiply a force.*

Figure I *A machine can also change speed.*

TRUE OR FALSE

In the space provided, write "true" if the sentence is true. Write "false" if the sentence is false.

_________ **1.** Your muscles do work.

_________ **2.** Muscles can do every job.

_________ **3.** Machines help our muscles.

_________ **4.** A complex machine is a combination of several simple machines.

_________ **5.** A slanted board is a machine.

_________ **6.** A wooden stick can be a machine.

_________ **7.** A machine can make a force stronger.

_________ **8.** One machine may be made of a combination of simple machines.

_________ **9.** A machine gets a job done with less work.

_________ **10.** A machine can only change the direction of a force.

FILL IN THE BLANK

Complete each statement using a term or terms from the list below. Write your answers in the spaces provided.

make a force stronger	inclined plane	screw
lever	machines	change speed
wedge	wheel and axle	pulley
six	work	change direction

1. ________________ help us do very hard jobs.
2. A machine can do one or more of the following: ________________ , ________________ , and ________________ .
3. There are only ________________ simple machines.
4. The names of the simple machines are the ________________ , the ________________ , the ________________ , the ________________ , the ________________ , and the ________________ .
5. Machines do not reduce the amount of ________________ .

WORD SEARCH

The list on the left contains words that you have used in this Lesson. Find and circle each word where it appears in the box. The spellings may go in any direction: up, down, left, right, or diagonally.

PULLEY
SCREW
WHEEL
AXLE
MACHINE
INCLINE
PLANE
LEVER
WEDGE

P	L	A	N	E	W	E	I
U	G	A	R	Y	S	D	N
L	E	E	H	W	C	G	C
L	M	I	L	S	R	E	L
E	N	I	H	C	A	M	I
Y	W	S	C	F	D	E	N
L	A	X	L	E	O	G	E
E	R	E	D	V	N	D	P
S	E	V	R	E	V	E	L
G	E	S	C	R	E	W	A

SCIENCE *EXTRA*

Machinists

In the early days of machine development, all machines were hand made one-at-a-time. Each part was custom fitted to match the other parts of the machine. The same parts of one machine were not interchangeable with the parts of another machine. Today, great numbers of identical machine parts are made. Any part can be replaced by another identical part. Interchangeable machine parts are the key to assembly line production. Making these parts requires a great deal of precision.

The precision parts of machines are made by machinists. Machinists are skilled workers who make and repair machines such as cars, aircraft, televisions, refrigerators, and other appliances. A machinist also may install or operate machinery.

Machinists work with metal. They follow blueprint plans and shape metals to specific sizes and shapes using power tools. Machinists bend, cut, drill, grind, hammer, and squeeze metal into any desired shape using these tools. Some machinists also use advanced machine tools which use electrical energy, chemical energy, heat, or magnetism to cut and shape extremely hard metals.

Are you handy? Do you like to work with tools? If so, you might wish to consider a career as a machinist. Skilled machinists are well paid and always in demand. To become a machinist, you need a high school diploma. Some vocational high schools offer machinist courses. After graduating from high school, several years of an apprenticeship is necessary. An apprentice works with an experienced machinist to learn and master the skills needed to be a machinist. After gaining the needed experience, many machinists work in production lines. Others set up their own shops.

What is mechanical advantage?

28

effort: the force that moves the resistance
mechanical advantage (M.A.): the number of times a machine multiples a force
resistance: the weight of an object to be moved

LESSON 28 What is mechanical advantage?

A machine can change a weak force into a strong force. We say a machine multiplies force. The number of times a machine multiplies a force is called its **mechanical advantage** or **M.A.**

Each machine has a special formula to find its M.A. But you can find the M.A. of any machine if you know:

- the **resistance** (the weight of the object to be moved)
- the **effort** (the force that moves the resistance)

To find the M.A. of any machine, you divide the force of the resistance by the force that moves the resistance. The formula is:

$$\text{M.A.} = \frac{\text{Resistance}}{\text{Effort}}$$

To understand how the formula works, let us look at two examples. Look at Figure A. An inclined plane lets you move a 10 Newton weight (resistance) with an effort force of 5 Newtons.

$$\text{M.A.} = \frac{\text{Resistance}}{\text{Effort}} = \frac{10}{5} = 2$$

The mechanical advantage is 2 (M.A. = 2).

This means that the inclined plane doubles the effort (5 × 2 = 10).

Look at Figure C. The effort is the same as the resistance. The mechanical advantage is one.

$$\text{M.A.} = \frac{\text{Resistance}}{\text{Effort}} = \frac{2}{2} = 1$$

Any number divided by the same number is 1.

There can be no zero mechanical advantage.

FINDING THE MECHANICAL ADVANTAGE

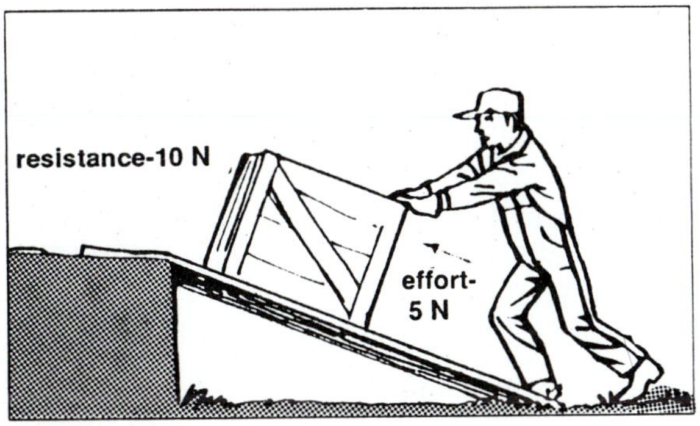

Figure A

1. The resistance force in Figure A is

 _______________ N.

2. The effort force is _______________ N.

3. The effort force is multiplied _________ times.

4. The M.A. is _______________ .

Figure B

5. The resistance force in Figure B is

 _______________ N.

6. The effort force is _______________ N.

7. The effort force is multiplied _________ times.

8. The M.A. is _______________ .

Figure C

9. The resistance force in Figure C is

 _______________ N.

10. The effort force is _______________ N.

11. Is this machine multiplying the effort

 force? _______________

12. The M.A. is _______________ .

13. What useful job is this machine doing?

 ____

 ____

MACHINES AND MECHANICAL ADVANTAGES

Different machines have different mechanical advantages. Remember no matter what the mechanical advantage is, the amount of work is the same.

Machines do not reduce the amount of work. They just make the work easier to do.

A machine divides a big job into several small jobs. Several small jobs are easier to do than one big job.

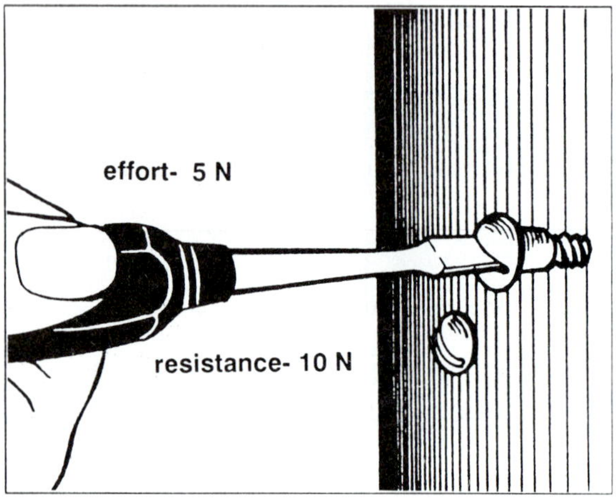

Figure D

1. The simple machine in Figure D is a

   ________________ .

2. The resistance is ________________ N.
3. The effort is ________________ N.
4. The effort is multiplied ________________ times.
5. The M.A. is ________________ .

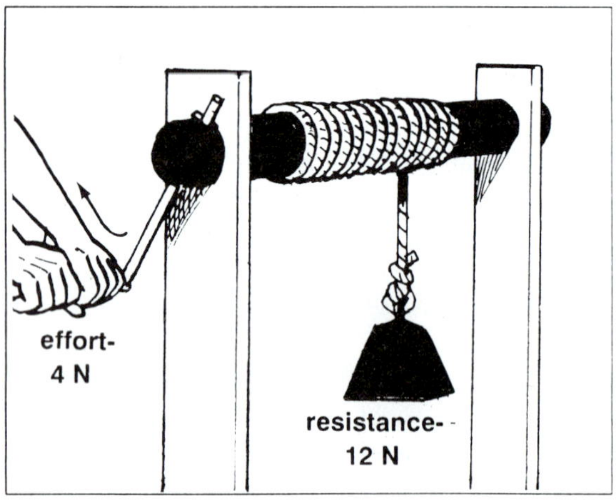

Figure E

6. The simple machine in Figure E is a

   ________________ .

7. The resistance is ________________ N.
8. The effort is ________________ N.
9. The effort is multiplied ________________ times.
10. The M.A. is ________________ .

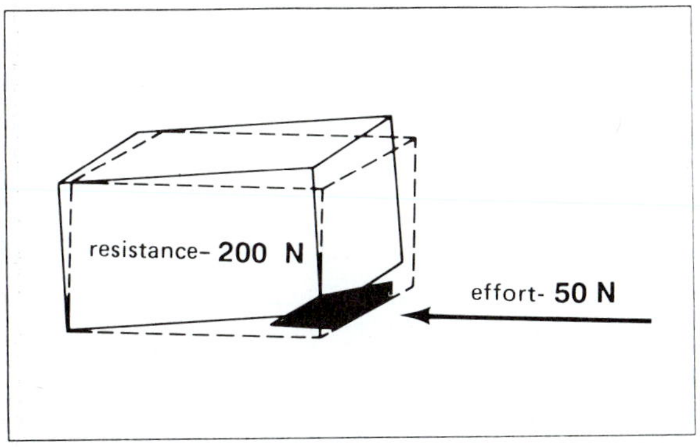

Figure F

11. The machine in Figure F is a

_________________ .

12. The resistance is _________________ N.

13. The effort is _________________ N.

14. The effort is multiplied _________________ times.

15. The M.A. is _________________ .

FILL IN THE BLANK

Complete each statement using a term or terms from the list below. Write your answers in the spaces provided.

effort	greater	multiplies
mechanical advantage	smaller and easier jobs	reduce work
less	resistance	easier
stronger		

1. Another name for any weight we try to move is _________________ .
2. The greater the weight, the _________________ the resistance.
3. The lighter the weight, the _________________ the resistance.
4. An _________________ is needed to move a resistance.
5. A machine makes a force _________________ .
6. Another way of saying this is, "A machine _________________ a force."
7. The number of times a machine multiplies a force is called its _________________ .
8. Machines do not _________________ .
9. Machines just make work _________________ to handle.
10. Machines divide work into _________________ .

COMPLETE THE CHART

Complete the chart by filling in the missing information.

$$\text{Remember M.A.} = \frac{\text{Resistance}}{\text{Effort}}$$

	Effort	Resistance	M.A.
1.	30 N	300 N	10
2.	5 N	25 N	
3.	15 N	15 N	
4.	20 N		2
5.	2 N	100 N	
6.		200 N	4
7.	25 N	150 N	
8.		150 N	5
9.	50 N	50 N	
10.	30 N		6

REACHING OUT

Solve these word problems. (Hint: Figure out which numbers you are given. Solve the problem for the missing number.)

1. A 200 N force moves an 800 N resistance. Find the M.A. ________________

2. A pulley lifts a 100 N resistance with a force of 100 N. Find the M.A. ________________

3. The M.A. of a certain machine is 2. How much force does it use to move a 200 N resistance?

____

4. The M.A. of a certain machine is 4. It lifts a resistance with 10 N. What is the weight of the resistance? ________________

What price do you pay to gain mechanical advantage?

29

effort distance: the distance an effort must be applied to move a resistance
resistance distance: the distance a resistance moves

LESSON 29 | What price do you pay to gain mechanical advantage?

There is a saying "You don't get anything for nothing." This may not be true for everything, but it is true for machines.

A machine does not multiply a force without "charging a price." The "price" is a greater distance the effort moves. When a machine multiplies a force, it also increases the effort distance.

To look at it another way, the less effort needed to do work, the longer the distance over which the effort must be spread. This means that the greater the mechanical advantage, the greater the distance the effort force must move.

There is a definite relationship between the **effort distance** and the **resistance distance**. For example, look at Figure A. This pulley has a mechanical advantage of 2. Using this pulley, you can lift a 10 Newton resistance with a 5 Newton effort force.

When you multiply the effort force, you must increase the effort distance.

You must pull the rope 2 meters to move the resistance only 1 meter.

You have learned that a machine does not reduce the amount of work. A machine reduces the amount of effort to do work.

A machine divides a big job into smaller, easier jobs. The smaller jobs are spread out over a longer distance. If you add up the work done in all the small jobs, it is the same amount of work used to do one big job.

RESISTANCE AND EFFORT DISTANCES

Figure A

The mechanical advantage of the pulley in Figure A is 2.

1. This pulley system multiplies the effort force ______________ times.
2. This pulley system multiplies the effort distance ______________ times.

How far the resistance moves is called the resistance distance.

How far you must pull the ropes is called the effort distance.

Figure out the effort distances needed, using the pulley in Figure A, to lift a resistance the distances listed below in Table 1. Then in Table 2, figure out the resistance distances using the effort distances given. The first one on both tables has been done for you.

Table 1

	Resistance Distance	Effort Distance
1.	2 meters	4 meters
2.	5 meters	
3.	10 meters	
4.	15 meters	
5.	20 meters	
6.	25 meters	

Table 2

	Resistance Distance	Effort Distance
7.	3 meters	6 meters
8.		8 meters
9.		12 meters
10.		16 meters
11.		40 meters
12.		60 meters

Figure B

Look at Figure B. The mechanical advantage of the pulleys in Figure B is 3. Using this pulley, you can lift a 9 Newton resistance with a 3 Newton effort force. You must pull the rope 3 meters to move the object only 1 meter.

3. How many times do these pulleys multiply an effort force? ________

4. How many times do they multiply a distance? ________

Using these pulleys, figure effort distances.

	Resistance Distance	Effort Distance
1.	2 meters	6 meters
2.	5 meters	
3.	10 meters	
4.	15 meters	
5.	20 meters	
6.	25 meters	

Now, figure resistance distances.

	Resistance Distance	Effort Distance
7.	2 meters	6 meters
8.		12 meters
9.		15 meters
10.		39 meters
11.		81 meters
12.		99 meters

TRUE OR FALSE

In the space provided, write "true" if the sentence is true. Write "false" if the sentence is false.

__________ 1. A machine makes work easier to do.

__________ 2. A machine reduces the amount of work that is done.

__________ 3. A machine increases the amount of work that is done.

__________ 4. A stick can be used as a machine.

__________ 5. A hill can be used as a machine.

__________ 6. A machine divides a small job into several larger jobs.

__________ 7. Mechanical advantage reduces effort.

__________ 8. Mechanical advantage reduces distance.

__________ 9. The greater the mechanical advantage, the easier the job becomes.

__________ 10. The greater the mechanical advantage a pulley has, the less rope you need.

FILL IN THE BLANK

Complete each statement using a term or terms from the list below. Write your answers in the spaces provided.

longer	20 meters	decrease
mechanical advantage	distance	10 meters
increases		

1. When a machine multiplies a force, it also multiplies ________________ .
2. The weaker the effort needed to do work, the ________________ the distance over which the work is spread.
3. The greater the ________________ , the greater the distance.
4. If the M.A. is 2 for a pulley, you must pull the rope ________________ to move the resistance 10 meters.
5. If the M.A. is 4, you must pull the rope 40 meters to move the resistance ________________ .
6. A simple machine ________________ the amount of effort.
7. Simple machines ________________ the distance over which the work is spread.

REACHING OUT

Figure C

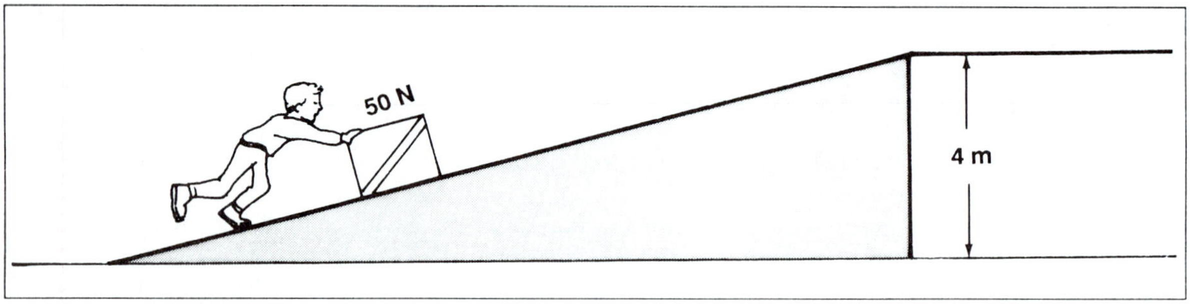

Figure D

Figures C and D show two inclined planes. A 50 N resistance is being pushed up each incline. The boys that are pushing the resistances are equally strong.

Incline C is shorter than incline D.

You would think that the boy using incline C would reach the top sooner than the boy using incline D. But this may not be so. The boy using incline D may reach the top before the boy using incline C.

Can you explain why? ______________________________

____

____

____

____

What is an inclined plane?

30

angle: the shape made by two straight lines that meet at a point
inclined plane: a slanted flat surface; a simple machine
slope: the height of a triangle divided by the length of its base

LESSON 30 | What is an inclined plane?

What is an **inclined plane**?

- A plane is a flat surface.
- An incline is a slant.
- An inclined plane, then, is a slanted flat surface.

The inclined plane was probably the first simple machine ever used. Early cave people must have found it easier to roll heavy stones up a hill than to carry them.

Think of the things you do. Which is easier, walking up steep steps or up a gentle ramp? Would you rather lift a heavy box or slide it up a slanted board?

An inclined plane multiplies force. It lets us move heavy objects with little force.

How much an inclined plane multiplies a force depends upon two things: .

- the height of the plane
- the length of the plane

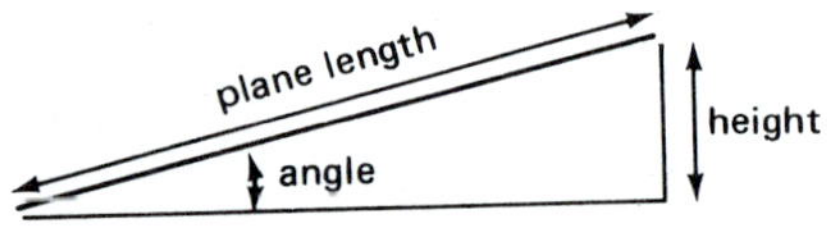

The length of the plane compared to its height gives a ramp its **slope** and **angle**.

An inclined plane may slope at any angle. Some slopes are gentle. Some are steep. The steeper the slope, the greater the angle.

This is an example of an inclined plane with a gentle slope.

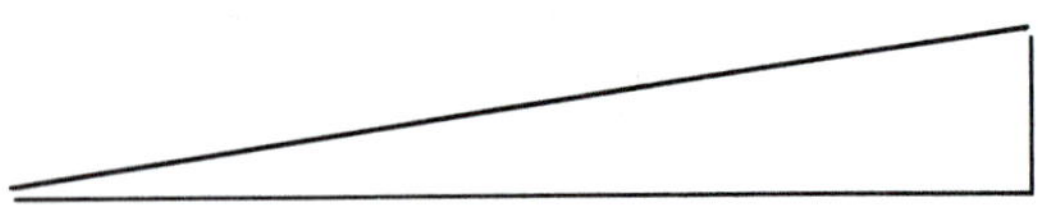

This is an example of an inclined plane with a steep slope.

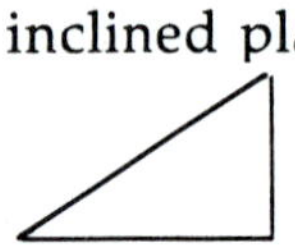

Study Figures A, B, and C. You will see that work is easier on a gentle slope than on a steep slope.

HOW DOES AN INCLINED PLANE MAKE WORK EASIER

Suppose that you have a crate that weighs 200 Newtons. You want to raise it onto a platform 2 meters high. How much force is needed? It depends. . .

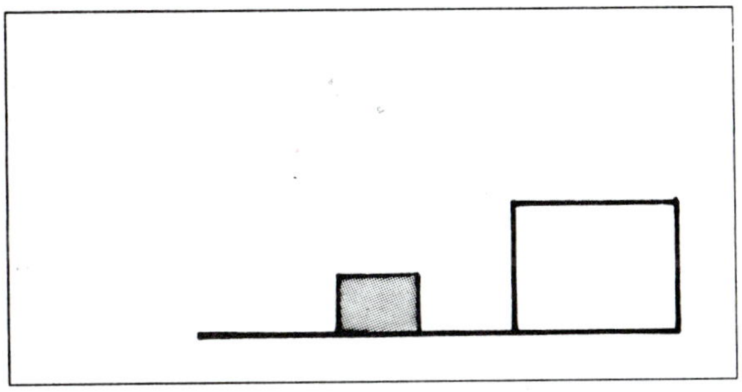

Figure A

WITHOUT A MACHINE you would need a force equal to the resistance: 200 N.

WITH A RAMP (INCLINED PLANE) the force needed depends upon the length (and the resulting slope) of the ramp.

Figure B

If the ramp is 4 meters long:

$$MA = \frac{4}{2} = 2$$

This ramp multiplies a force 2 times. Therefore, a force of 100 Newtons is needed.

Figure C

If the ramp is 8 meters long:

$$MA = \frac{8}{2} = 4$$

This ramp multiplies a force 4 times. Therefore, a force of 50 Newtons is needed.

Figure D

If the ramp is 16 meters long:

$$MA = \frac{16}{2} = 8$$

This ramp multiplies a force 8 times. Therefore, a force of 25 Newtons is needed.

Yes, an inclined plane makes work easier to do. it multiplies force. But it also multiplies distance. The weaker the force, the longer the distance over which it must be used. In the end the <u>work</u> is the same. It makes no difference if you lift a resistance or use a ramp.

Now answer these questions by indicating Figure B, C, or D. Fill in the blanks for the other sentences.

1. Which ramp multiplies force the most? ______________
2. Which ramp multiplies force the least? ______________
3. Which ramp is the longest? ______________
4. Which ramp is the shortest? ______________
5. Which ramp has the steepest slope? ______________
6. Which ramp has the most gentle slope? ______________
7. The ramp with steepest slope has the ______________ M.A.
   largest, smallest
8. The ramp with the gentlest slope has the ______________ M.A.
   largest, smallest
9. The ramp with the steepest slope has the ______________ distance to the top.
   longest, shortest
10. The ramp with the gentlest slope has the ______________ distance to the top.
    longest, shortest

FILL IN THE BLANK

Complete each statement using a term or terms from the list below. Write your answers in the spaces provided. Some answers may be used more than once.

mechanical advantage	work	force
sloping flat surface	flat surface	length
height		

1. A plane is a ______________ .
2. An inclined plane is a ______________.
3. An inclined plane multiplies a ______________ .
4. The number of times an inclined plane multiplies a force is called its

   ______________ .
5. The M.A. of an inclined plane depends upon two things. They are the ______________ of

   the plane and its ______________.
6. An inclined plane does not reduce ______________.
7. An inclined plane lets you do work with less ______________.

FIGURING MECHANICAL ADVANTAGE

Can you divide four by two, or nine by three? Of course you can! Then you can figure the mechanical advantage of an inclined plane.

To find the M.A. of an inclined plane, just divide the length of the incline by the height.

$$\text{M.A.} = \frac{\text{Length of incline}}{\text{Height}}$$

Let's try an example.

This inclined plane is 6 meters long and 3 meters high.

$$\text{M.A.} = \frac{\text{Length of incline}}{\text{Height}}$$

$$\text{M.A.} = \frac{6}{3} = 2$$

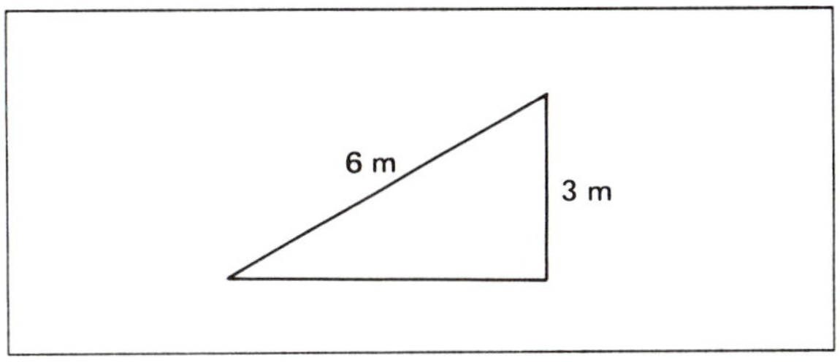

Figure E

This means that the incline multiplies a force by two. A resistance can be pushed up this incline with a force one-half its weight.

For example, a 200 Newton resistance can be pushed with a force of 100 N.

1. In Figure F, the length of the inclined plane is ____________ .
2. The height is ____________ .
3. The mechanical advantage is ____________ .
4. A force is multiplied ____________ .

How much effort force will move each of the resistances given in the table on the right? The first one has been done for you.

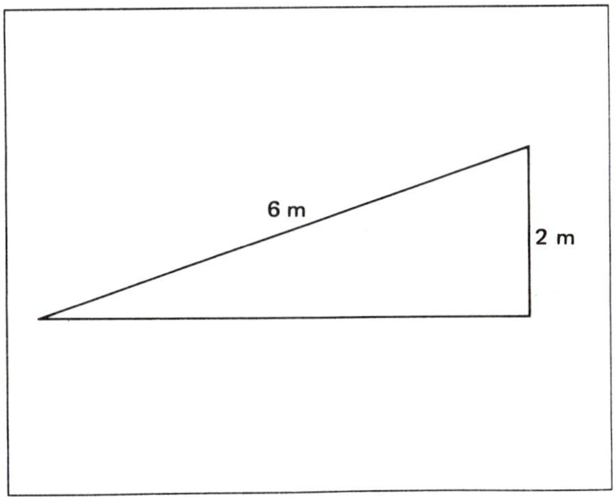

Figure F

	Resistance	Effort Force
5.	15 N	5 N
6.	30 N	
7.	39 N	
8.	120 N	
9.	150 N	
10.	180 N	

COMPLETE THE CHART

Complete the chart by filling in the missing information.

	Length of Incline	Height	M.A.	Number of Times the Force Is Multiplied	Resistance	Force
1.	10 meters	5 meters			100 N	
2.	50 meters	10 meters				30 N
3.	30 meters	6 meters			60 N	
4.	90 meters	30 meters			75 N	
5.	100 meters	25 meters				100 N

MULTIPLE CHOICE

In the space provided, write the letter of the word that best completes each statement.

_______ **1.** The number of times a machine multiplies a force is called its
a) mechanical ability **b)** mechanical disadvantage
c) mechanical advantage **d)** kinetic energy

_______ **2.** A mother is pushing a baby carriage. The carriage is the
a) effort force **b)** resistance
c) work **d)** mechanical advantage

_______ **3.** In pushing the baby carriage, the mother's pushing is the
a) effort force **b)** resistance
c) work **d)** mechanical advantage

_______ **4.** A plane is
a) an angle **b)** curved
c) a simple machine **d)** flat

_______ **5.** Another word for incline is
a) plane **b)** curve
c) angle **d)** height

_______ **6.** An incline is
a) straight with the ground **b)** at an angle with the ground
c) always steep **d)** always flat

How is a wedge like an inclined plane?

31

wedge: a sloping solid that comes to a point; a simple machine

LESSON 31 How is a wedge like an inclined plane?

Could you break a log into two pieces with your bare hands? Of course not. You could, however, if you used a wedge.

To split a log, you place a wedge into a narrow crack of a log. Then, hit the wedge hard with a long hammer. After several blows, the log will split in two.

The wedge multiplied the force of each blow. It made a hard job easier to do.

What is a wedge? How does it multiply a force?

A **wedge** is a sloping solid that comes to a point. So is an inclined plane. How are they different? The difference can be described in just one word, <u>movement</u>.

An inclined plane does not move, A wedge does its job by moving.

A wedge can work in any direction. Let us study a downward-moving wedge, such as the one shown in Figure A.

When a downward force hits against a wedge, the point moves downward. As the point moves downward, the sides push outward. The outward force is very great. It can split objects.

A wedge can also lift heavy objects. it can even prevent objects from moving. A wedge stops movement when it is forced between an object that we want to stop moving, and an object that <u>cannot</u> move, like a wall or floor.

A wedge can have any number of slopes. Some have one slope. Some have two slopes. Some wedges are sloped all around.

A wedge is like any other simple machine. It does not save work. It just makes work easier to handle.

HOW A WEDGE WORKS

Figure A shows a wedge with a downward force.

A force moves the point downward. As the point moves downward, the sides push outward.

The outward force is very great. It is the downward force multiplied several times.

Figure A

WEDGE SLOPES

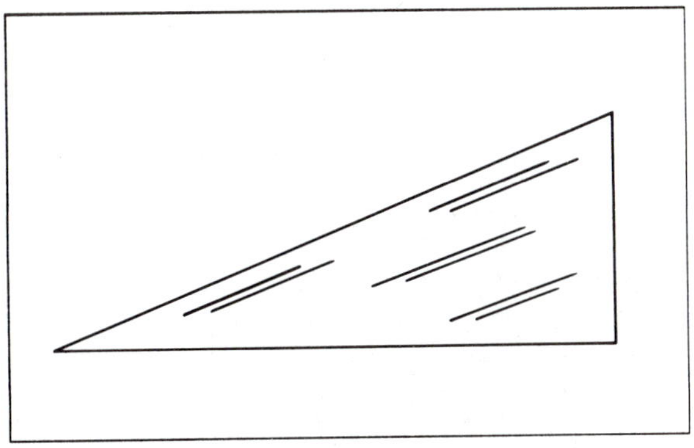

Figure B *Some wedges have one slope.*

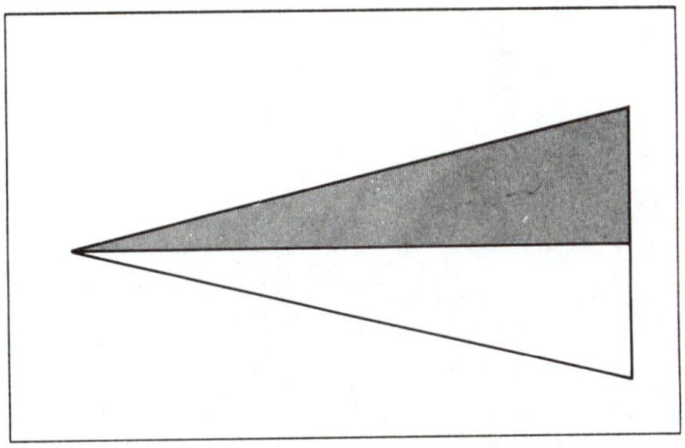

Figure C *Some wedges have two slopes. This is like two inclined planes back to back.*

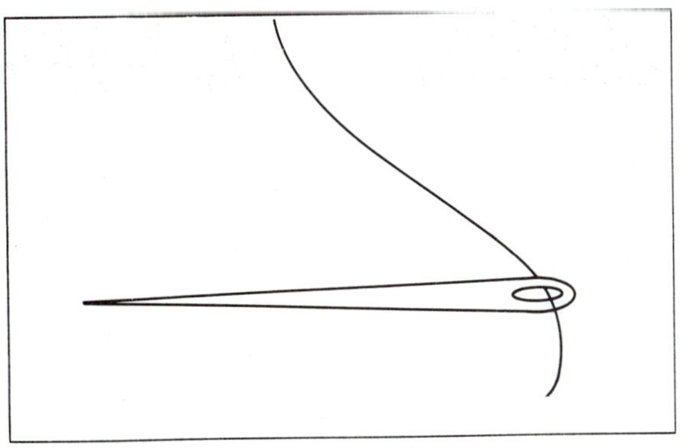

Figure D *Some wedges are sloped all around.*

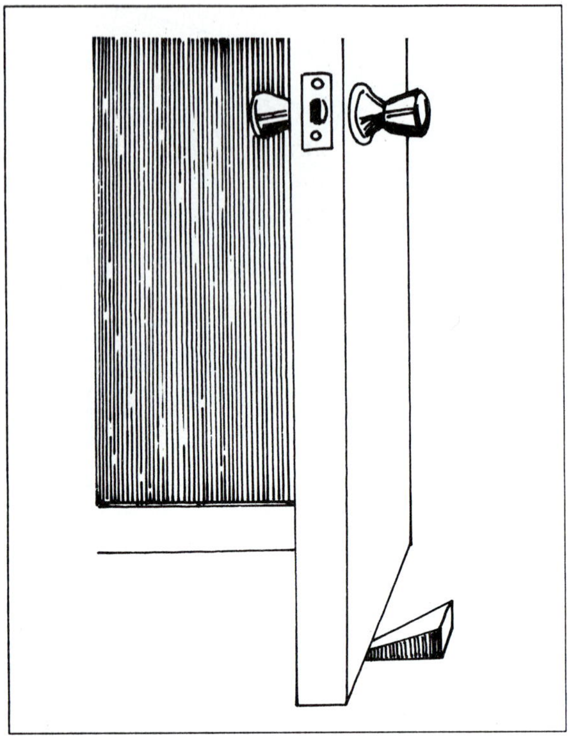

Figure E

Figure F

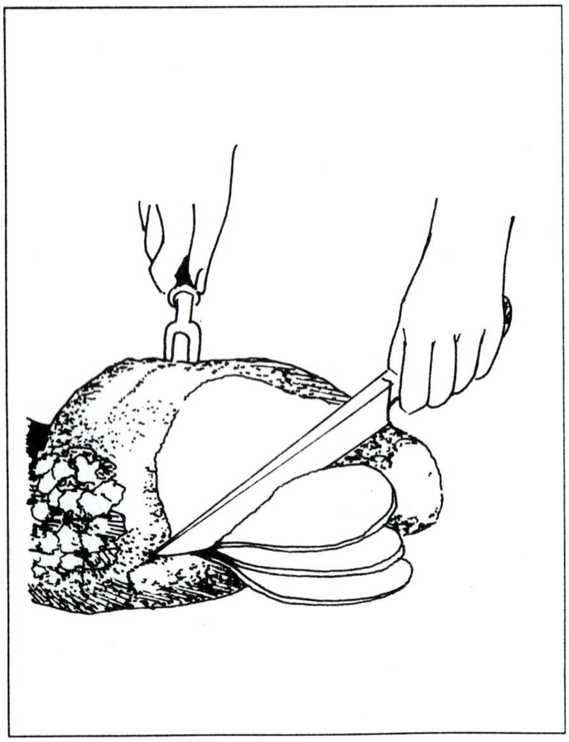

Figure G

Figure H

FILL IN THE BLANK

Complete each statement using a term or terms from the list below. Write your answers in the spaces provided.

back to back	simple machines	point
sloping solid	distance	inclined planes
easier	reduce	stronger
any number	moves	does not move
the same		

1. A wedge is one of the six ________________ .
2. A machine can make a force ________________ .
3. A machine does not ________________ work.
4. A machine makes work ________________ to handle.
5. A machine multiplies a force. But it also multiplies ________________ .
6. With or without a machine, the work is ________________ .
7. A wedge is a ________________ that comes to a ________________ .
8. The difference between an inclined plane and a wedge is that an inclined plane ________________ . A wedge ________________ .
9. A wedge may have ________________ of slopes.
10. A wedge with two slopes is like two ________________ that are ________________ .

DRAWING WEDGES

1. Draw a wedge with a single, gentle slope.

Figure F

2. Draw a wedge with a single, steep slope.

Figure G

MATCHING

Match each term in Column A with its description in Column B. Write the correct letter in the space provided.

	Column A	Column B
_______	1. to make work easier	a) jobs of a wedge
_______	2. moves objects and prevents objects from moving	b) does not move
_______	3. inclined plane	c) the number of times a force is multiplied
_______	4. wedge	d) main job of machines
_______	5. mechanical advantage	e) helps work by moving

WORD SEARCH

The list on the left contains words that you have used in this Lesson. Find and circle each word where it appears in the box. The spellings may go in any direction: up, down, left, right, or diagonally.

simple machine
wedge
screw
inclined plane
pulley
lever
wheel and axle
force
work
slope
angle
resistance
effort

S	E	V	Y	D	E	T	A	R	U	A	L	I
L	I	E	O	A	B	D	S	E	V	Y	G	N
O	Y	M	R	N	D	O	E	S	R	E	O	C
P	G	O	P	G	F	O	R	C	E	B	O	L
E	W	O	H	L	E	V	E	R	F	D	S	I
F	I	N	B	E	E	F	I	E	F	D	E	N
E	H	O	U	S	B	M	N	W	O	R	K	E
N	I	N	T	R	E	O	A	S	R	S	T	D
Y	E	L	L	U	P	F	T	C	T	K	Z	P
C	G	E	M	U	H	E	N	T	H	R	E	L
N	D	A	E	V	F	N	U	A	R	I	T	A
R	E	S	I	S	T	A	N	C	E	Q	N	N
X	W	H	E	E	L	A	N	D	A	X	L	E

How is a screw like an inclined plane?

32

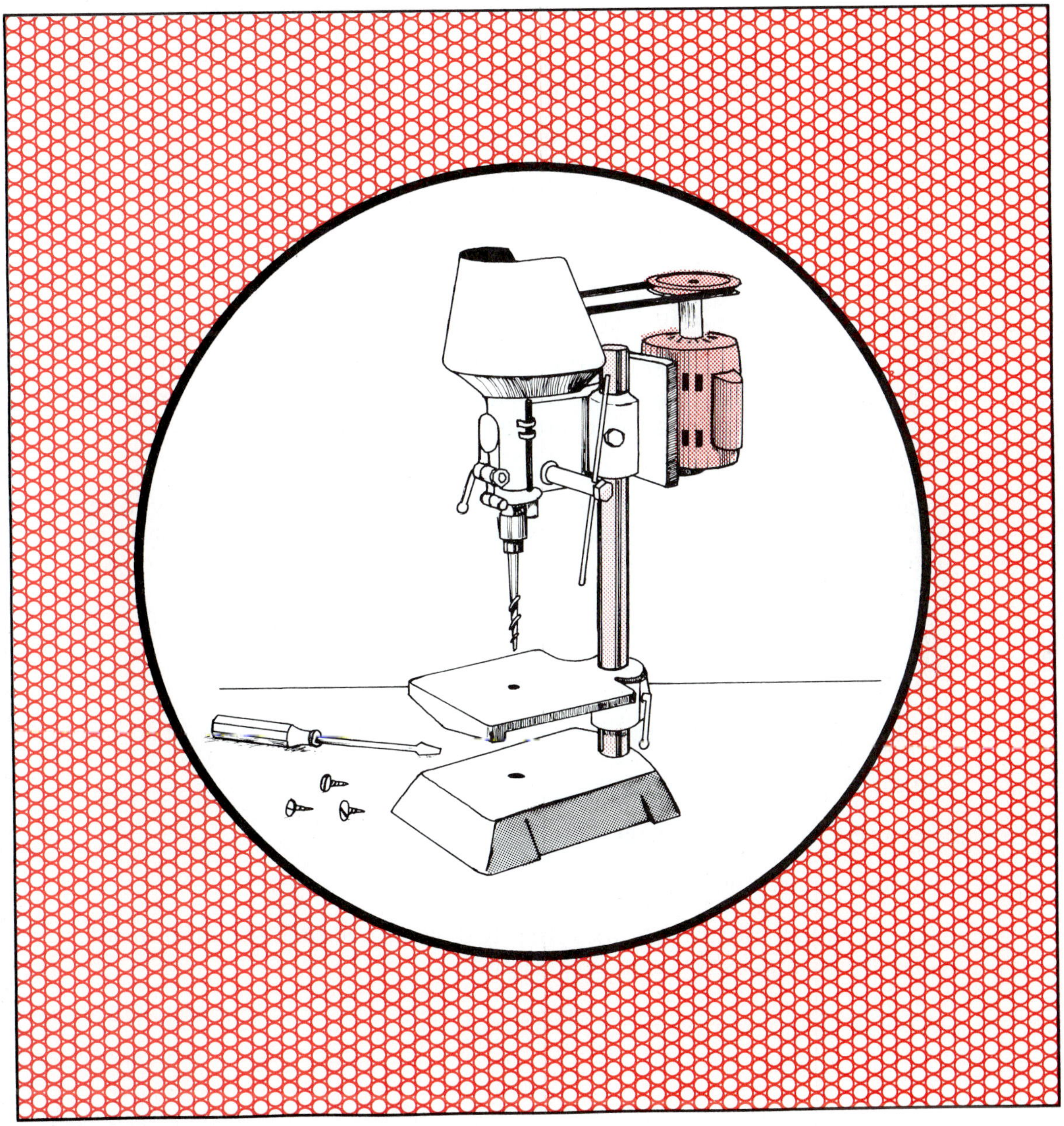

grooves: the low points on a screw
ridges: the high points on a screw; threads
screw: an inclined plane that spirals around a metal rod; a simple machine
threads: the high points on a screw; ridges

LESSON 32 How is a screw like an inclined plane?

Look at Figure A and try this simple experiment:

- Cut a piece of paper in the shape of a long triangle (Step 1).
- Now wrap it around a pencil (Step 2).
- Examine the paper carefully (Step 3).

Notice that the paper winds around the pencil at an angle. What does it look like. Doesn't it remind you of a screw?

What you have made is really the idea behind the screw. A **screw** is an inclined plane that spirals around a metal rod. The winding forms **grooves** (low points), and **ridges** (high points) (Figure B). The ridges are also called **threads**.

If you could unwind the thread of a screw, you would have a very long inclined plane. It is especially long compared to the height of its rod. This great difference in size gives a screw a very large mechanical advantage. Remember,

$$\text{M.A.} = \frac{\text{Length of incline}}{\text{Height}}.$$

The longer the incline compared to the height, the greater the M.A.

The screw helps us in two ways: .

- A screw holds things together.
- A screw raises heavy objects.

A screw is like any simple machine. It does not reduce the amount of work. It just makes work easier to handle.

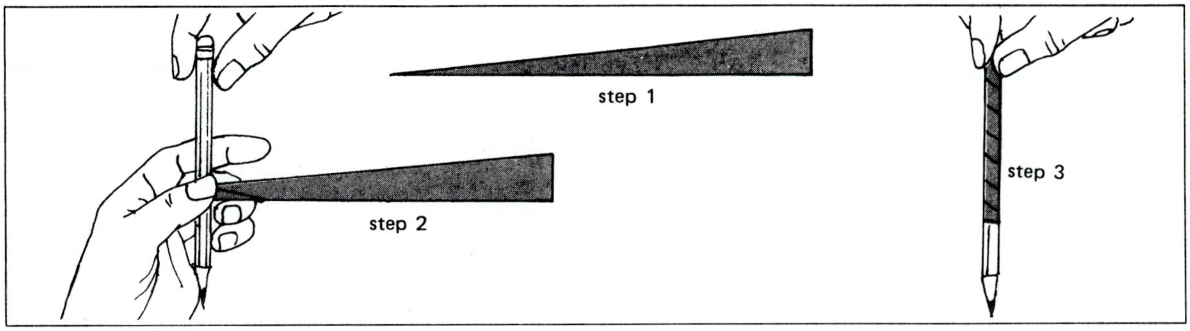

Figure A *The wound paper shows grooves and ridges. So does a screw.*

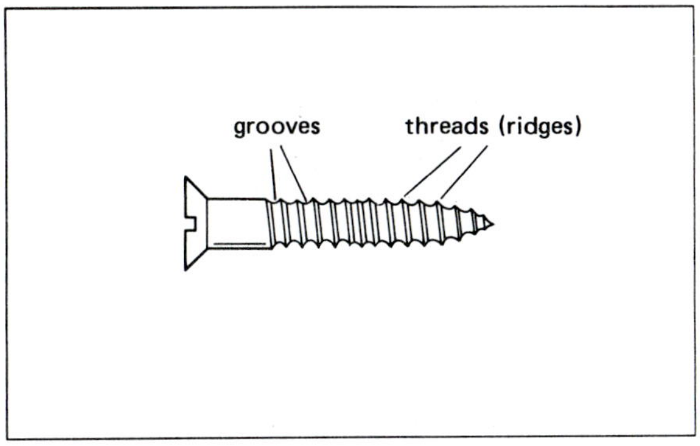

Figure B *A household screw.*

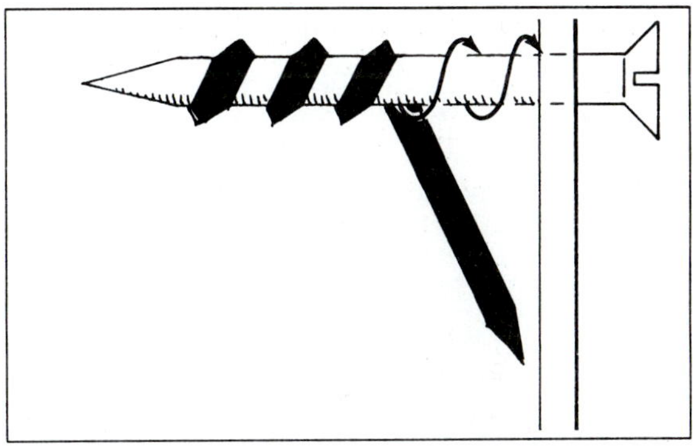

Figure C *If a screw could be "unwound," we would see a long inclined plane.*

Figure D *A jack screw*

A screw multiplies a force. It also multiplies distance. Take the jack screw in Figure D for example. You must turn the screw a lot in order to raise the resistance a little bit.

FILL IN THE BLANK

Complete each statement using a term or terms from the list below. Write your answers in the spaces provided.

easier to handle	ridges	very long
multiplies force	an inclined plane	lift
metal rod	fasten	threads
work	simple machines	force
mechanical advantage	distance	

1. A screw is one of the six ________________ .
2. A screw is a machine because it ________________ .
3. A screw is really ________________ wound around a ________________ .
4. The spiral of a screw usually is ________________ compared to its height.
5. A long spiral gives a screw a large ________________ .
6. The high points of a screw are called the ________________ or ________________ .
7. Screws do two kinds of jobs. They ________________ and ________________ .
8. A screw is like any other machine. It multiplies ________________ as well as ________________ .
9. A machine does not reduce the amount of ________________ .
10. A machine makes work ________________ .

MATCHING

Match each term in Column A with its description in Column B. Write the correct letter in the space provided.

	Column A	Column B
________	1. machines	a) a simple machine
________	2. screw	b) high points of a screw
________	3. grooves	c) make work easier
________	4. ridges	d) large M.A.
________	5. long incline compared to height	e) low points of a screw

STUDYING SCREWS

Two screws are shown in Figures E and F. The lengths are the same. But the threads are different. Study each screw. Then answer the questions and fill in the blanks.

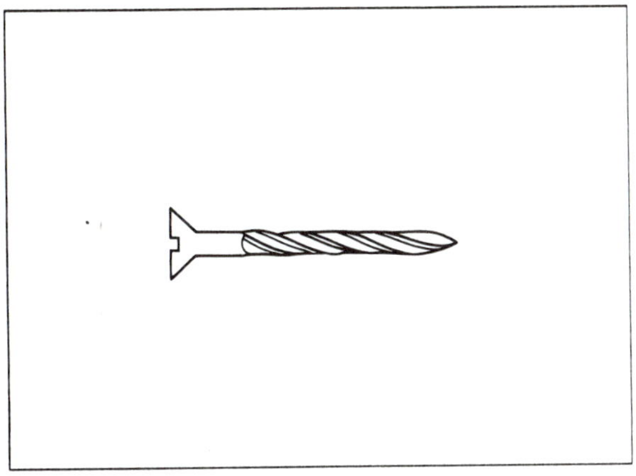

Figure E

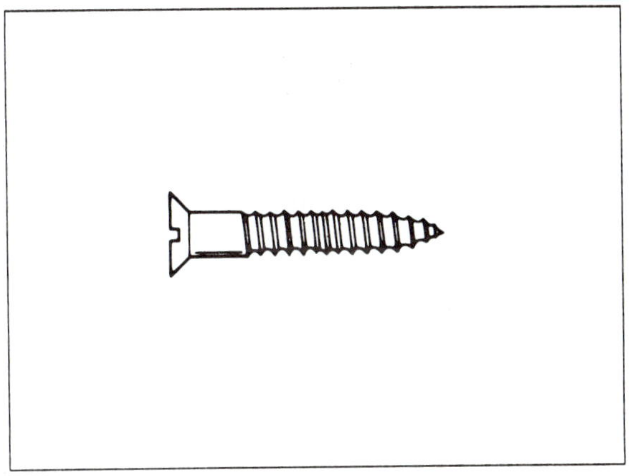

Figure F

1. Which screw. . .

   a) has more threads? ____________________

   b) has fewer threads? ____________________

   c) has the larger M.A.? ____________________

   d) has the smaller M.A.? ____________________

   e) will take more turns to attach? ____________________

   f) will take fewer turns to attach? ____________________

2. Which thread angle . . .

   a) is steeper? ____________________

   b) is less steep? ____________________

3. The steeper the thread angle, the ____________________ (longer, shorter) the inclined plane.

4. The more gentle the thread angle, the ____________________ (longer, shorter) the inclined plane.

5. The steeper the thread angle, the ____________________ (greater, less) the M.A.

6. The more gentle the thread angle, the ____________________ (greater, less) the M.A.

TRUE OR FALSE

In the space provided, write "true" if the sentence is true. Write "false" if the sentence is false.

_________ **1.** A screw is like a winding inclined plane.

_________ **2.** A screw does its job by moving.

_________ **3.** All screws give the same M.A.

_________ **4.** The thread of every screw is longer than its height.

_________ **5.** Screws only hold things together.

_________ **6.** Screws can be used to move things.

_________ **7.** Ridges is another word for threads.

_________ **8.** A screw reduces the amount of work.

_________ **9.** A screw with more threads has a smaller M.A.

_________ **10.** A screw multiplies force.

WORD SCRAMBLE

Below are several scrambled words you have used in this Lesson. Unscramble the words and write your answers in the spaces provided.

1. DHEART _________________

2. VORGOE _________________

3. WRECS _________________

4. RIPLAS _________________

5. NICLEIN _________________

What is a lever?

33

effort arm: the distance between a fulcrum and an effort force
fulcrum [FUL-krum]: the steady point on which the plank or rod of a lever rests
lever: a rod or plank which rests on and can be rotated on a fulcrum; a simple machine
resistance arm: the distance between a fulcrum and a resistance

LESSON 33 | What is a lever?

Did you ever pry a lid from a can? If you did, you have used a lever.

A **lever** can multiply a force many times. It lets you do tough jobs that you cannot do with your muscles alone. A lever also changes the direction of a force.

A lever has two main parts:

- a rod or plank that is free to turn
- a steady point for the plank to rest and turn upon. We call this steady point the **fulcrum** [FUL-krum].

A resistance rests on one side of the fulcrum. An effort presses on the other side. The effort is the force needed to move the resistance.

The distance between the fulcrum and resistance is called the **resistance arm** or resistance distance.

The distance between the fulcrum and effort is called the **effort arm** or effort distance.

You can change the resistance distance and effort distance by changing the fulcrum. When the fulcrum changes, the M.A. changes too.

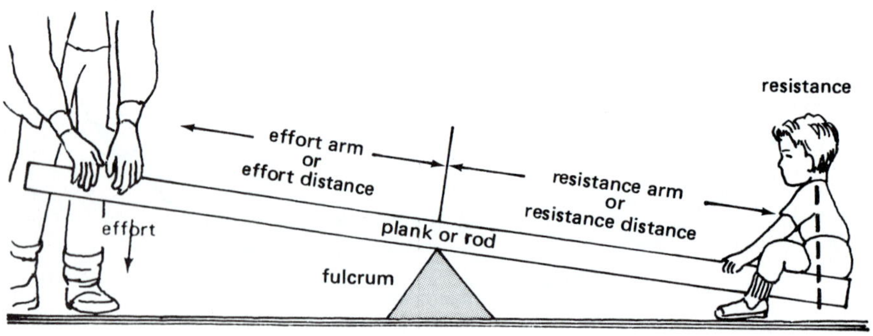

MAKE YOUR OWN LEVER

You can make a simple lever with just two pencils. Place a book on one end of a pencil. Press on the other end with one finger. The lever lets you lift the book easily.

See what happens when you move the fulcrum.

Figure A

Place the fulcrum far away from the book as shown in Figure A. Press down on the effort arm. Notice that it is not easy to lift the book. A strong effort is needed.

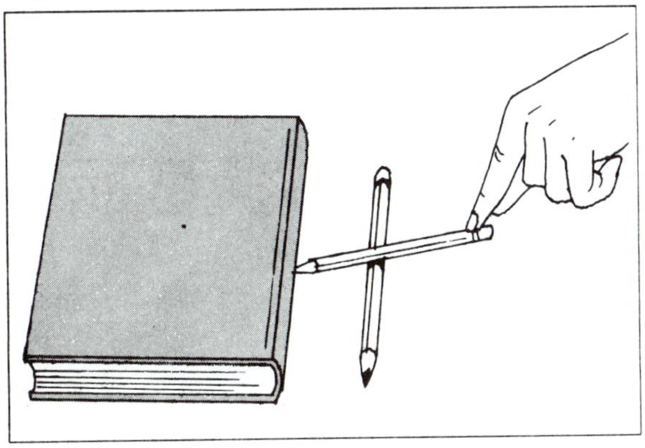

Figure B

Move the fulcrum closer to the book—just under the middle of the pencil as shown in Figure B. Press down on the effort arm. Is it easier or more difficult to move the book?

Figure C

Move the fulcrum right up against the book. Press down on the effort arm. Is it easier or more difficult to move the book, than when the fulcrum was under the middle of the pencil?

We see that work becomes easier as the effort arm becomes longer. Another way of saying this is the longer the effort arm compared to the resistance arm, the greater the mechanical advantage.

You can easily figure the mechanical advantage of a lever. Just divide the effort arm length by the resistance arm length. Here is an example:

If an effort arm is 4 centimeters long and a resistance arm is 2 centimeters long, then the mechanical advantage is 2.

$$\text{M.A.} = \frac{\text{Effort arm}}{\text{Resistance arm}} = \frac{4\text{ cm}}{2\text{ cm}} = 2$$

1. In Figure A, the longer arm is the ____________ .
effort arm, resistance arm

2. Compared to the resistance arm, the effort arm is ____________ .
longer, shorter

3. In Figure B, the effort arm has been made ____________ than in Figure A.
longer, shorter

4. To move the resistance, you now need ____________ effort.
more, less

5. The longer effort arm made the job ____________ .
easier, more difficult

6. In Figure C, the effort arm has been made even ____________ than in figure B.
longer, shorter

7. To move the resistance, you now need even ____________ force.
more, less

8. The M.A. is ____________ .
increasing, decreasing

9. The longer the effort arm, the ____________ the M.A.
larger, smaller

10. The shorter the effort arm, the ____________ the M.A.
larger, smaller

FINDING THE M.A.

Two small levers are shown in Figures D and E. Study each one. Then answer the questions. Be careful, one is tricky!

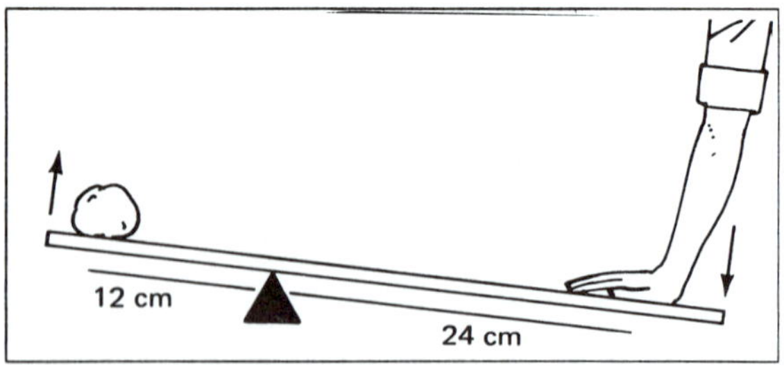

Figure D

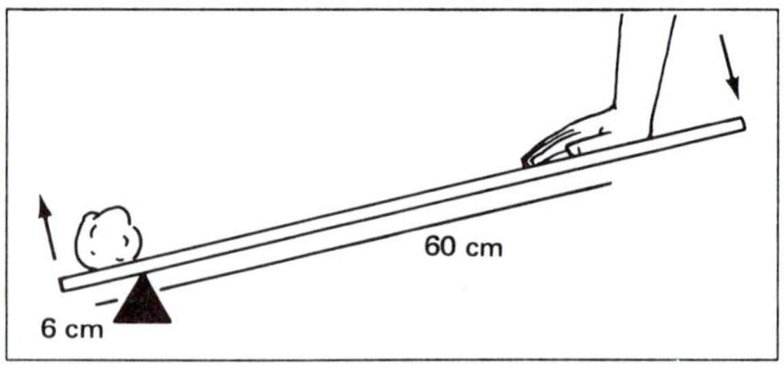

Figure E

1. $\frac{\text{Effort arm}}{\text{Resistance arm}}$ = ________ = M.A.

2. A force is multiplied ________ times.

3. A 50 Newton effort force will lift a resistance of ________ N.

4. $\frac{\text{Effort arm}}{\text{Resistance arm}}$ = ________ = M.A.

5. A force is multiplied ________ times.

6. A 10 Newton effort force will lift ________ N.

FILL IN THE BLANK

Complete each statement using a term or terms from the list below. Write your answers in the spaces provided. Some words may be used more than once.

lever	turns	effort distance
effort	multiply	fulcrum
changes the direction	rod	resistance
resistance distance	plank	

1. A see-saw is an example of a ________________.
2. The main parts of a lever are a ________________ point and a ________________ or ________________.
3. The steady point that a lever plank rests upon is called the ________________.
4. The rod or plank ________________ on the fulcrum.
5. A ________________ rests on one side of the fulcrum; an ________________ presses on the other side.
6. The distance between the fulcrum and the resistance is called the ________________.
7. The distance between the fulcrum and the effort is called the ________________.
8. When you change the position of a ________________, you change the effort distance and the resistance distance.
9. When you change the effort distance and resistance distance, you change the position of the ________________.
10. A lever does two jobs, it can ________________ a force. It also ________________ of a force.

MATCHING

Match each term on Column A with its description in Column B. Write the correct letter in the space provided.

	Column A	Column B
______	1. lever	a) parts of a lever
______	2. fulcrum and plank	b) distance from fulcrum to resistance
______	3. effort	c) steady point for a plank to turn or rest
______	4. resistance arm	d) a simple machine
______	5. fulcrum	e) overcomes resistance

TRUE OR FALSE

In the space provided, write "true" if the sentence is true. Write "false" if the sentence is false.

_________ **1.** A lever is a complicated machine.

_________ **2.** A lever moves straight with the ground.

_________ **3.** A lever rotates on its effort arm.

_________ **4.** Every lever has a fulcrum.

_________ **5.** The fulcrum of every lever is at the same spot.

_________ **6.** You can change the M.A. of a lever by changing its fulcrum.

_________ **7.** You can change the M.A. of a lever by changing its effort distance.

_________ **8.** You can change the M.A. of a lever by changing its resistance distance.

_________ **9.** A long effort arm always means a large M.A. (Think carefully about this one).

_________ **10.** The longer the effort arm compared to the resistance arm, the greater the M.A.

REACHING OUT

Why must long levers be made stronger than short levers?

____

____

____

____

____

____

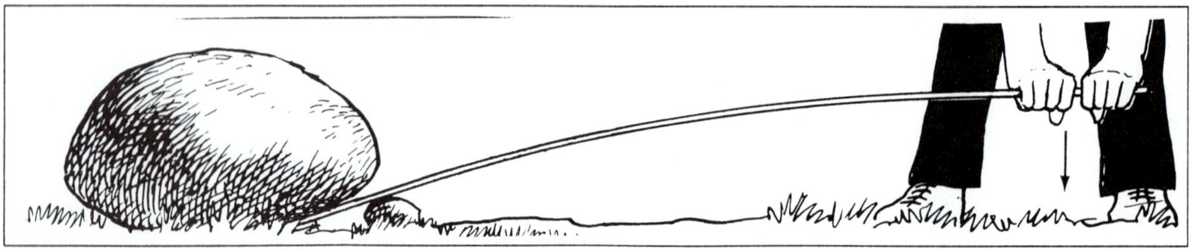

Figure F

What are a wheel and axle?

34

axle: a thin shaft or rod
diameter [dy-AM-ut-ur]: a line that passes through the center of a circle
wheel and axle: a wheel with a rod attached to its center; a simple machine

LESSON 34 | What are a wheel and axle?

Every time you turn a doorknob, you use a machine. A doorknob is part of a **wheel and axle**. The knob is the wheel; the thin shaft connected to the knob is the shaft or **axle** (Figure A).

The knob makes it easy to turn the axle.

Think of your own experiences. Did you ever try to open a door that had a missing knob? Was it easy or difficult to turn the shaft?

A wheel and axle multiplies a force. It builds a small force into a larger force.

The amount a wheel and axle multiplies a force depends upon two things:

- the diameter of the wheel
- the diameter of the axle

A **diameter** [dy-AM-ut-ur] is a line that passes through the center of a circle.

How do you find the mechanical advantage of a wheel and axle? it is easy. Just divide the diameter of the wheel by the diameter of the axle.

Look at Figure B for example. The diameter of the wheel is 40 centimeters. The axle is 10 centimeters in diameter. The mechanical advantage, then, is 4.

$$\text{M.A.} = \frac{\text{Diameter of wheel}}{\text{Diameter of axle}}$$

$$\text{M.A.} = \frac{40\text{ cm}}{10\text{ cm}} = 4 \qquad \text{M.A.} = 4$$

A wheel and axle is like any other simple machine. What it adds in M.A., it loses in the distance the effort has to move.

The wheel and axle is also like any simple machine. It does not reduce the amount of work. It just makes work easier to do.

WHICH IS WHICH?

Figure A

1. A complete door knob is a simple machine called a ______________ .
2. a) the wheel is labeled ______________ .

   b) the axle is labeled ______________ .
3. The diameter of the wheel is ______________ (larger, smaller) than the diameter of the axle.
4. A wheel and axle is a machine because it ______________ a force.
5. A machine may multiply a force. But it also multiplies ______________ .

FINDING MECHANICAL ADVANTAGE

A wheel and axle is like any other machine. It multiplies force. But, it also multiplies distance.

Look at the wheel and axle in Figure B, for example. It multiplies a force 4 times. But to raise a resistance just 20 centimeters, you must turn the large wheel 80 centimeters.

The mechanical advantage of this wheel and axle is 4.

1. Keeping the axle at 10 cm, what would the M.A. be with a

   a) 30 cm wheel? ______________

   b) 50 cm wheel? ______________
2. Keeping the wheel at 40 cm, what would the mechanical advantage be with a

   a) 5 cm axle? ______________

   b) 20 cm axle? ______________

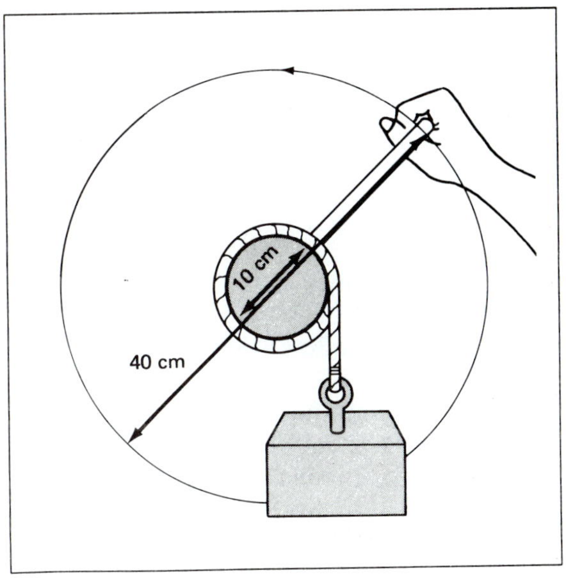

Figure B

WHAT IS A DIAMETER?

A diameter is a line that passes through the center of a circle. It passes from one part of the circle to the opposite part.

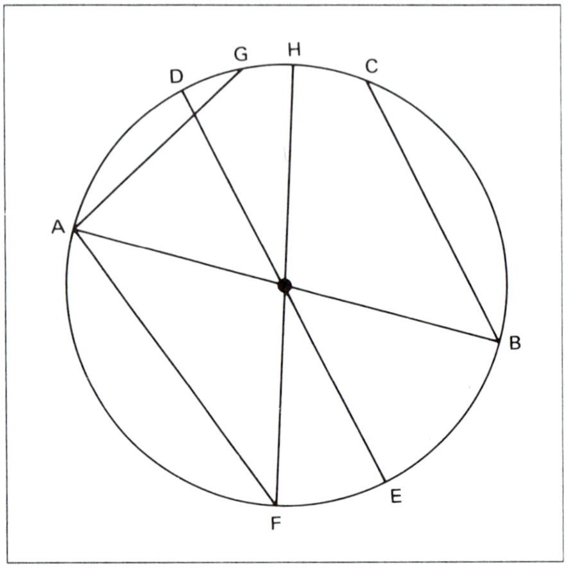

Figure C

Some of the lines within the circle in Figure A are diameters. Some are not.

Study the lines. Then put a check (✓) in the proper box.

	Line	Is a Diameter	Is Not a Diameter
1.	AB		
2.	AG		
3.	BC		
4.	FH		
5.	AF		
6.	DE		

Now fill in the answer to these questions.

1. Measure the lines that are diameters. How long is each one? ____________
2. Are they all the same length? ____________
3. In any one circle, the diameters are the ____________ length.

MEASURING DIAMETERS

Four wheels and axles are shown below.

Measure the wheel and axle of each one with your metric ruler. (Use the mm measurements).

Then figure the M.A. for each.

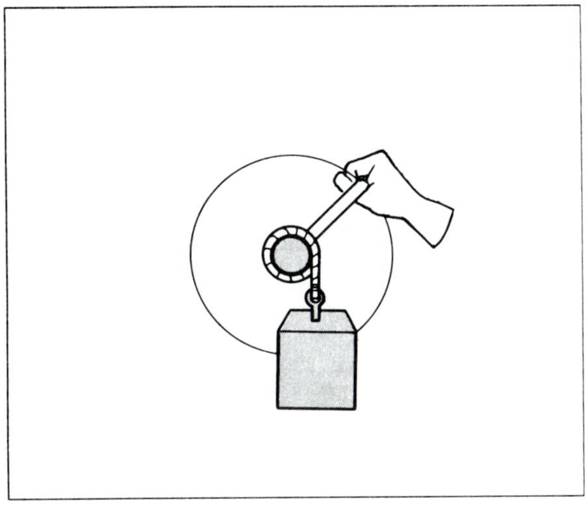

Figure D

1. Wheel diameter = ____________ mm

   Axle diameter = ____________ mm

   M.A. = ____________

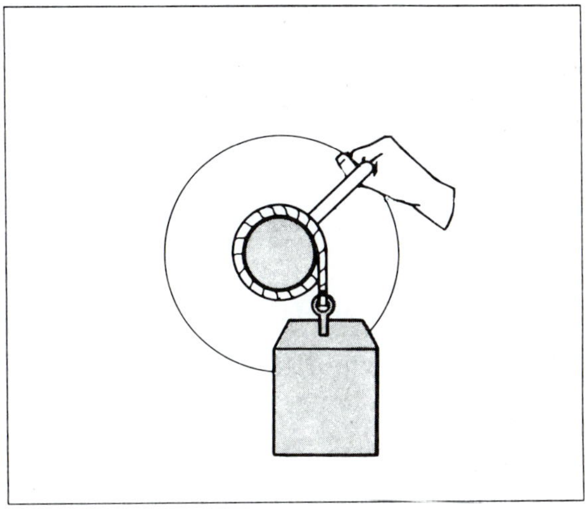

Figure E

2. Wheel diameter = ____________ mm

   Axle diameter = ____________ mm

   M.A. = ____________

Figure F

1. Wheel diameter = ____________ mm

   Axle diameter = ____________ mm

   M.A. = ____________

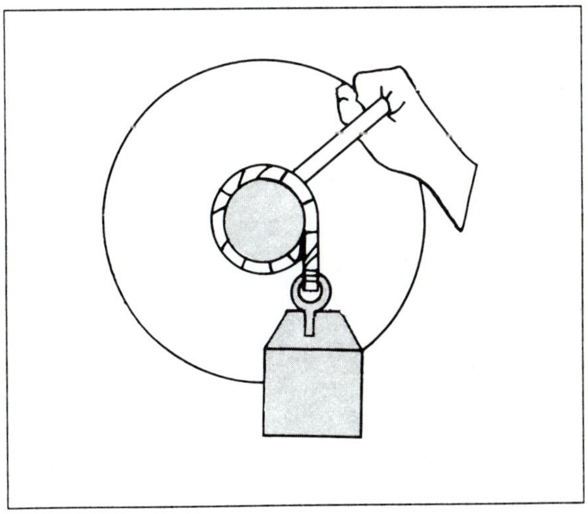

Figure G

2. Wheel diameter = ____________ mm

   Axle diameter = ____________ mm

   M.A. = ____________

FINDING THE M.A. OF WHEELS AND AXLES

Find the mechanical advantage of each of these wheels and axles. Then fill in the blanks below.

	Wheel Diameter	Axle Diameter	M.A.
1.	20 cm	2 cm	
2.	22 cm	2 cm	
3.	26 cm	2 cm	
4.	30 cm	2 cm	
5.	38 cm	2 cm	

6. The axle diameters ______________ (are, are not) the same.

7. The wheel diameters ______________ (are, are not) the same.

8. The largest M.A. is the one with the ______________ (largest, smallest) wheel.

9. The smallest M.A. is the one with the ______________ (largest, smallest) wheel.

10. To gain more M.A. you use a larger ______________ (axle, wheel) .

11. If you were to use a thicker axle, the M.A. would become ______________ (larger, smaller) .

12. If you were to use a thinner axle, the M.A. would become ______________ (larger, smaller) .

REACHING OUT

A wheel and axle is a simple machine. But it has another simple machine built-in.

Which one is it? ______________________________

What is a pulley?

35

block and tackle: a system of pulleys
pulley: a grooved wheel that turns on a rod called an axle
single fixed pulley: a pulley attached to something that does not move
single moveable pulley: a pulley attached to an object which is being moved

LESSON 35 | What is a pulley?

Did you ever raise or lower a window blind? If you have, you have used a pulley. When you raise a blind, does the cord move up or down? When you lower a blind, does the cord move up or down?

A **pulley** is a grooved wheel that turns on a rod called an axle. Pulleys can do two kinds of jobs.

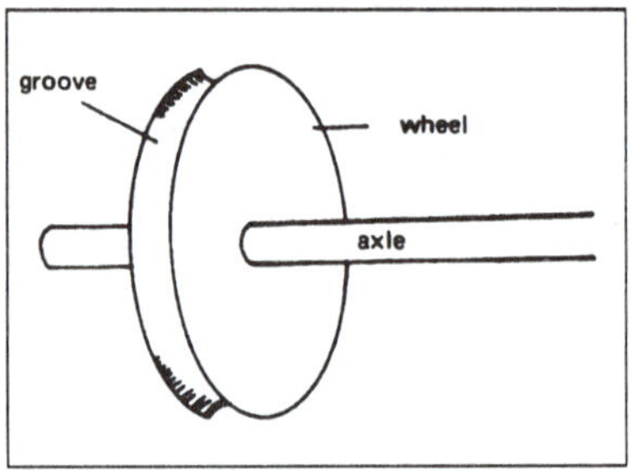

- Pulleys can change the direction of the effort.
- Pulleys can also multiply the effort force.

There are three kinds of pulleys:

- The **single fixed pulley** does not multiply force. It has an M.A. of one. A single fixed pulley just changes the direction of a force.
- The **single movable pulley** multiplies a force by two (M.A. = 2). It doubles a force. But it does not change the direction of a force.
- The **block and tackle** is a system of pulleys. Some pulleys in the system move. Some do not. A block and tackle multiplies a force many times. It is used to raise very heavy objects. A block and tackle also changes the direction of a force.

It is easy to figure the mechanical advantage of a pulley. Just count the number of strands holding up the weight. This number is the same as the mechanical advantage. For example, if 3 strands are holding up a resistance, then the M.A. is 3.

When you count the pulley strands, you <u>do not</u> count the strand that is being pulled. There is one exception. On a single movable pulley, you count all the strands. (There are two).

Pulleys that multiply force do not reduce the amount of work. They just make work easier to handle.

FIXED AND MOVEABLE PULLEYS

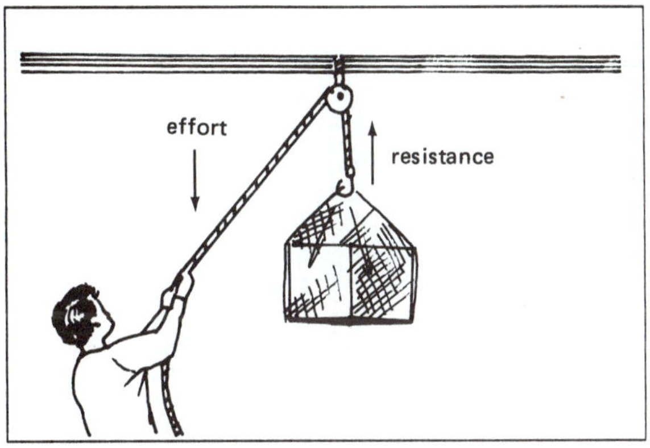

Figure A *Fixed pulley*

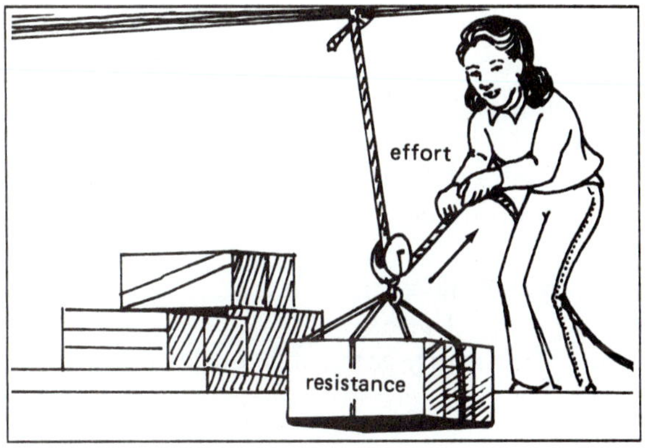

Figure B *Moveable pulley*

1. The pulley in Figure A ________________ (does, does not) move.

2. A pulley that does not move is called a ________________ pulley.

3. How many strands are holding up the resistance? ________________

4. The M.A. is ________________ .

5. A single fixed pulley ________________ (does, does not) multiply a force.

6. As the effort moves down, the resistance moves ________________ (up, down) .

7. Using this pulley, you need a force of ________________ to lift a 50 Newton resistance.

8. The pulley in Figure B ________________ (does, does not) move.

9. A pulley that moves is called a ________________ pulley.

10. How many strands are holding up the resistance? ________________

11. The M.A. is ________________ .

12. A single moveable pulley multiplies a force ________________ times.

13. As the effort moves up, the resistance moves ________________ (up, down) .

14. A single moveable pulley ________________ (does, does not) change direction.

15. Using this pulley, you need a force of ________________ to lift a 50 Newton resistance.

BLOCK AND TACKLE

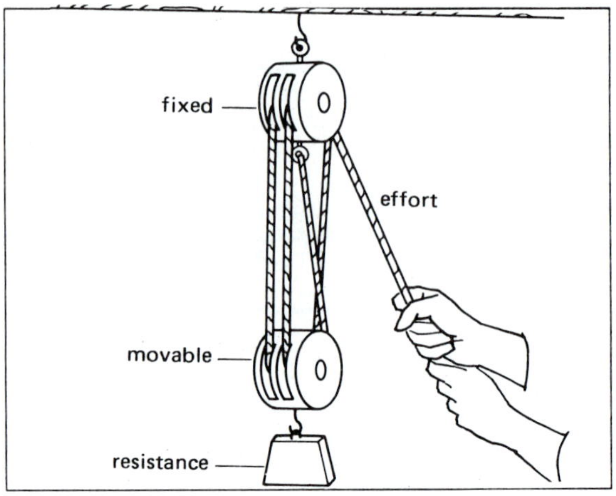

Figure C

1. Figure C shows a pulley system. It is called a ________________ .
2. Each wheel is considered a pulley. How many fixed pulleys does this block and tackle have? ________________
3. How many moveable pulleys does it have? ________________
4. How many strands are holding up the resistance? ________________
5. The M.A. is ________________ .
6. This block and tackle multiplies a force ________________ times.
7. It ________________ change the direction of the force.
   does, does not
8. Using this block and tackle, how much force would you need to lift each of these resistances?

   a) 100 Newtons ________________ b) 200 Newtons ________________

   c) 400 Newtons ________________ d) 1000 Newtons ________________

PULLEYS MULTIPLY DISTANCE

Figure D

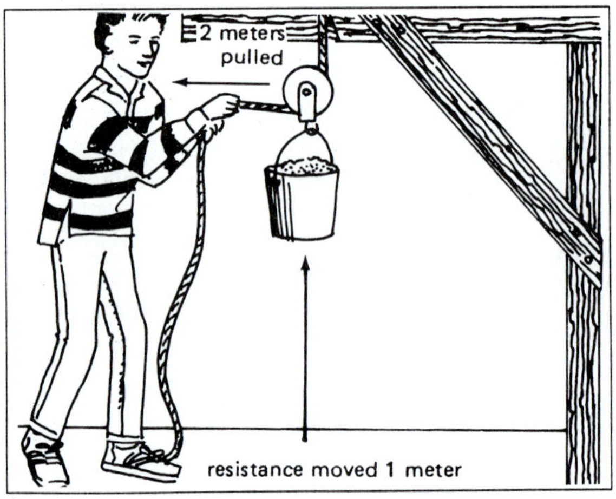

Figure E

A pulley is like any other machine. It multiplies force, as well as distance. Take a single moveable pulley for example (M.A. = 2). For every one meter you lift the resistance, you must pull the rope two meters. In the end, the work done is the same, with or without the pulley.

FIGURING MECHANICAL ADVANTAGE

What is the M.A. of each of these pulleys or pulley systems?

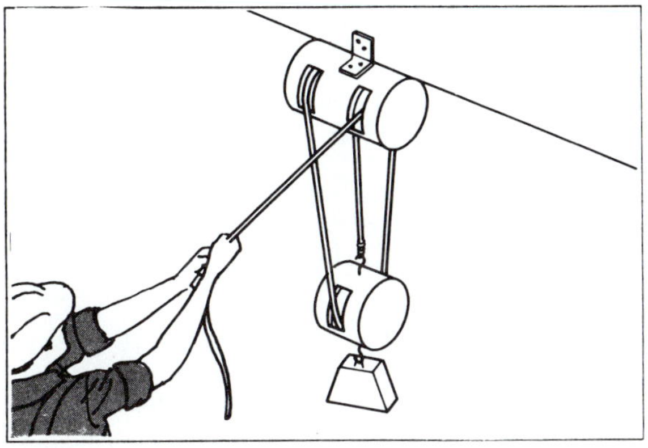

Figure F

1. M.A. = ____________________

Figure G

2. M.A. = ____________________

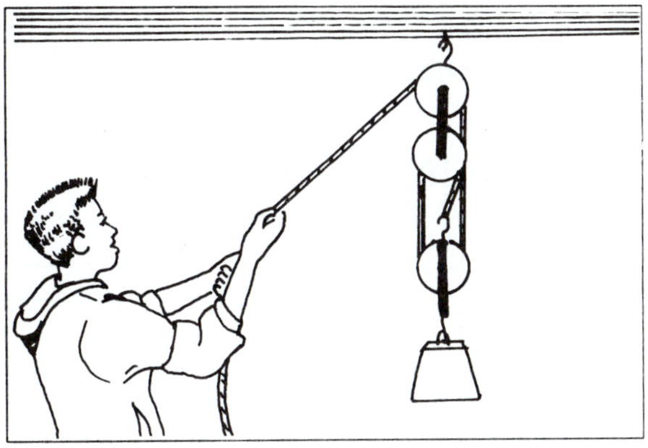

Figure H

3. M.A. – ____________________

Figure I

4. M.A. = ____________________

MATCHING

Match each term in Column A with its description in Column B. Write the correct letter in the space provided. Column A

	Column A	Column B
______	**1.** single moveable pulley	**a)** the number of strands pulling the object
______	**2.** block and tackle	**b)** used to lift very heavy objects
______	**3.** single fixed pulley	**c)** M.A. of 2
______	**4.** M.A. of most pulleys	**d)** only changes the direction of a force
______	**5.** pulley	**e)** grooved wheel that turns on a rod

PERIODIC TABLE OF ELEMENTS

Key

Atomic number →	29	2
		8 ← *Number of electrons in each shell*
Symbol →	Cu	18
		1
Name →	Copper	
	63.54	← *Atomic mass*

Period	1	2		3	4	5	6	7	8	9
1	1 **H** Hydrogen 1.00 (1)									
2	3 **Li** Lithium 6.93 (2 1)	4 **Be** Beryllium 9.01 (2 2)								
3	11 **Na** Sodium 22.98 (2 8 1)	12 **Mg** Magnesium 24.31 (2 8 2)								
4	19 **K** Potassium 39.10 (2 8 8 1)	20 **Ca** Calcium 40.08 (2 8 8 2)		21 **Sc** Scandium 44.95 (2 8 9 2)	22 **Ti** Titanium 47.90 (2 8 10 2)	23 **V** Vanadium 50.94 (2 8 11 2)	24 **Cr** Chromium 51.99 (2 8 13 1)	25 **Mn** Manganese 54.93 (2 8 13 2)	26 **Fe** Iron 55.84 (2 8 14 2)	27 **Co** Cobalt 58.93 (2 8 15 2)
5	37 **Rb** Rubidium 85.47 (2 8 18 8 1)	38 **Sr** Strontium 87.62 (2 8 18 8 2)		39 **Y** Yttrium 88.90 (2 8 18 9 2)	40 **Zr** Zirconium 91.22 (2 8 18 10 2)	41 **Nb** Niobium 92.90 (2 8 18 12 1)	42 **Mo** Molybdenum 95.94 (2 8 18 13 1)	43 **Tc** Technetium (97) (2 8 18 13 2)	44 **Ru** Ruthenium 101.07 (2 8 18 15 1)	45 **Rh** Rhodium 102.90 (2 8 18 16 1)
6	55 **Cs** Cesium 132.90 (2 8 18 18 8 1)	56 **Ba** Barium 137.34 (2 8 18 18 8 2)	57-70 La Series	71 **Lu** Lutetium 174.97 (2 8 18 32 9 2)	72 **Hf** Hafnium 178.49 (2 8 18 32 10 2)	73 **Ta** Tantalum 180.94 (2 8 18 32 11 2)	74 **W** Tungsten 183.85 (2 8 18 32 12 2)	75 **Re** Rhenium 186.2 (2 8 18 32 13 2)	76 **Os** Osmium 190.2 (2 8 18 32 14 2)	77 **Ir** Iridium 192.2 (2 8 18 32 15 2)
7	87 **Fr** Francium (223) (2 8 18 32 18 8 1)	88 **Ra** Radium (226) (2 8 18 32 18 8 2)	89-102 Ac Series	103 **Lr** Lawrencium (257) (2 8 18 32 32 9 2)	104 *	105 *	106 *	107 *	108 *	109 *

Lanthanide Series	57 **La** Lanthanum 138.91 (2 8 18 18 9 2)	58 **Ce** Cerium 140.12 (2 8 18 19 9 2)	59 **Pr** Praseodymium 140.90 (2 8 18 20 9 2)	60 **Nd** Neodymium 144.24 (2 8 18 22 8 2)	61 **Pm** Promethium (145) (2 8 18 23 8 2)	62 **Sm** Samarium 150.35 (2 8 18 24 8 2)	63 **Eu** Europium 151.96 (2 8 18 25 8 2)
Actinide Series	89 **Ac** Actinium (227) (2 8 18 32 18 9 2)	90 **Th** Thorium 232.03 (2 8 18 32 19 9 2)	91 **Pa** Protactinium (231) (2 8 18 32 20 9 2)	92 **U** Uranium 238.03 (2 8 18 32 21 9 2)	93 **Np** Neptunium (237) (2 8 18 32 22 9 2)	94 **Pu** Plutonium (244) (2 8 18 32 23 9 2)	95 **Am** Americium (243) (2 8 18 32 24 9 2)

*** Names for these elements have not been agreed upon.**

10	11	12	13	14	15	16	17	18
								2 **He** Helium 4.00 (2)
			5 **B** Boron 10.81 (2 3)	6 **C** Carbon 12.01 (2 4)	7 **N** Nitrogen 14.00 (2 5)	8 **O** Oxygen 15.99 (2 6)	9 **F** Fluorine 18.99 (2 7)	10 **Ne** Neon 20.18 (2 8)
			13 **Al** Aluminum 26.98 (2 8 3)	14 **Si** Silicon 28.08 (2 8 4)	15 **P** Phosphorus 30.97 (2 8 5)	16 **S** Sulfur 32.06 (2 8 6)	17 **Cl** Chlorine 35.45 (2 8 7)	18 **Ar** Argon 39.94 (2 8 8)
28 **Ni** Nickel 58.71 (2 8 16 2)	29 **Cu** Copper 63.54 (2 8 18 1)	30 **Zn** Zinc 65.37 (2 8 18 2)	31 **Ga** Gallium 69.72 (2 8 18 3)	32 **Ge** Germanium 72.59 (2 8 18 4)	33 **As** Arsenic 74.92 (2 8 18 5)	34 **Se** Selenium 78.96 (2 8 18 6)	35 **Br** Bromine 79.90 (2 8 18 7)	36 **Kr** Krypton 83.80 (2 8 18 8)
46 **Pd** Palladium 106.4 (2 8 18 18 0)	47 **Ag** Silver 107.87 (2 8 18 18 1)	48 **Cd** Cadmium 112.40 (2 8 18 18 2)	49 **In** Indium 114.82 (2 8 18 18 3)	50 **Sn** Tin 118.69 (2 8 18 18 4)	51 **Sb** Antimony 121.75 (2 8 18 18 5)	52 **Te** Tellurium 127.60 (2 8 18 18 6)	53 **I** Iodine 126.90 (2 8 18 18 7)	54 **Xe** Xenon 131.30 (2 8 18 18 8)
78 **Pt** Platinum 195.09 (2 8 18 32 16 2)	79 **Au** Gold 196.96 (2 8 18 32 18 1)	80 **Hg** Mercury 200.59 (2 8 18 32 18 2)	81 **Tl** Thallium 204.37 (2 8 18 32 18 3)	82 **Pb** Lead 207.19 (2 8 18 32 18 4)	83 **Bi** Bismuth 208.98 (2 8 18 32 18 5)	84 **Po** Polonium (209) (2 8 18 32 18 6)	85 **At** Astatine (210) (2 8 18 32 18 7)	86 **Rn** Radon (222) (2 8 18 32 18 8)

64 **Gd** Gadolinium 157.25 (2 8 18 25 9 2)	65 **Tb** Terbium 158.92 (2 8 18 26 9 2)	66 **Dy** Dysprosium 162.50 (2 8 18 28 8 2)	67 **Ho** Holmium 164.93 (2 8 18 29 8 2)	68 **Er** Erbium 167.26 (2 8 18 30 8 2)	69 **Tm** Thulium 168.93 (2 8 18 31 8 2)	70 **Yb** Ytterbium 173.04 (2 8 18 32 8 2)
96 **Cm** Curium (247) (2 8 18 32 25 9 2)	97 **Bk** Berkelium (247) (2 8 18 32 26 9 2)	98 **Cf** Californium (251) (2 8 18 32 28 8 2)	99 **Es** Einsteinium (254) (2 8 18 32 29 8 2)	100 **Fm** Fermium (257) (2 8 18 32 30 8 2)	101 **Md** Mendelevium (256) (2 8 18 32 31 8 2)	102 **No** Nobelium (254) (2 8 18 32 32 8 2)

THE METRIC SYSTEM

METRIC-ENGLISH CONVERSIONS

	Metric to English	*English to Metric*
Length	1 kilometer = 0.621 mile (mi)	1 mi = 1.61 km
	1 meter = 3.28 feet (ft)	1 ft = 0.305 m
	1 centimeter = 0.394 inch (in)	1 in = 2.54 cm
Area	1 square meter = 10.763 square feet	1 ft^2 = 0.0929 m^2
	1 square centimeter = 0.155 square inch	1 in^2 = 6.452 cm^2
Volume	1 cubic meter = 35.315 cubic feet	1 ft^3 = 0.0283 m^3
	1 cubic centimeter = 0.0610 cubic inches	1 in^3 = 16.39 cm^3
	1 liter = .2642 gallon (gal)	1 gal = 3.79 L
	1 liter = 1.06 quart (qt)	1 qt = 0.94 L
Mass	1 kilogram = 2.205 pound (lb)	1 lb = 0.4536 kg
	1 gram = 0.0353 ounce (oz)	1 oz = 28.35 g
Temperature	Celsius = 5/9 (°F −32)	Fahrenheit = 9/5°C + 32
	0°C = 32°F (Freezing point of water)	72°F = 22°C (Room temperature)
	100°C = 212°F (Boiling point of water)	98.6°F = 37°C (Human body temperature)

METRIC UNITS

The basic unit is printed in capital letters.

Length	*Symbol*
Kilometer	km
METER	m
centimeter	cm
millimeter	mm
Area	*Symbol*
square kilometer	km^2
SQUARE METER	m^2
square millimeter	mm^2
Volume	*Symbol*
CUBIC METER	m^3
cubic milliliter	mm^3
liter	L
milliliter	mL
Mass	*Symbol*
KILOGRAM	kg
gram	g
Temperature	*Symbol*
degree Celsius	°C

SOME COMMON METRIC PREFIXES

Prefix		*Meaning*
micro-	=	0.000001, or 1/1,000,000
milli-	=	0.001, or 1/1000
centi-	=	0.01, or 1/100
deci-	=	0.1, or 1/10
deka-	=	10
hecto-	=	100
kilo-	=	1000
mega-	=	1,000,000

SOME METRIC RELATIONSHIPS

Unit	*Relationship*
kilometer	1 km = 1000 m
meter	1 m = 100 cm
centimeter	1 cm = 10 mm
millimeter	1 mm = 0.1 cm
liter	1 L = 1000 mL
milliliter	1 mL = 0.001 L
tonne	1 t = 1000 kg
kilogram	1 kg = 1000 g
gram	1 g = 1000 mg
centigram	1 cg = 10 mg
milligram	1 mg = 0.001 g

GLOSSARY/INDEX

machine: any device that can change the strength of a force, its speed, or its direction, 170

malleable [MAL-ee-uh-bul]: able to be hammered into different shapes, 86

manometer [muh-NAHM-uh-tur]: an instrument that measures water pressure, 156

mass: amount of matter in an object, 2, 20

matter: anything that has mass and takes up space, 20

mechanical advantage (M.A.): the number of times a machine multiplies a force, 176

metallic luster: shine of a metal, 86

metals: elements found on the left side of the periodic table, which share properties, 86, 80

neutron [NEW-trahn]: a part of the atom that has neither a positive or negative charge; is found in the nucleus, 56

Newtons: units used to measure force, 116, 154

nonmetals: elements found on the right side of the periodic table, 80, 92

nucleus [NEW-klee-us]: central part of an atom, which contains neutrons and protons, 56

period: a row across the periodic table, 80

periodic [pir-ee-AHD-ik] **table:** a chart of all known elements, 80

potential [puh-TEN-shul] **energy:** stored energy, 104

pressure: a force on a definite area, 154

property [PROP-ur-tee]: characteristic used to describe a substance, 20

proton [PROH-tahn]: a part of the atom that has a positive charge; is found in the nucleus, 56

pulley: a grooved wheel that turns on a rod called an axle, 170, 218

resistance: the weight of an object to be moved, 176

resistance arm: the distance between a fulcrum and a resistance, 206

resistance distance: the distance a resistance moves,182, 206

resultant: final amount and direction of a force, 124

ridges: the high points on a screw; threads, 200

safety alert symbols: signs that warn of hazards or dangers, 16

scientific method: problem solving guide, 10

screw: an inclined plane that spirals around a metal rod; a simple machine, 170, 200

shells: energy levels in which electrons are arranged around the nucleus, 62

simple machines: basic machines; inclined plane, wedge, screw, lever, wheel and axle, and pulley, 170

single fixed pulley: a pulley attached to something that does not move, 218

single moveable pulley: a pulley attached to an object which is being moved, 218

slope: the height of a triangle divided by the length of its base, 194

solid: phase of matter with a definite shape and a definite volume, 26

states: phases of matter, 26

symbol: short way of writing the name of an element, 74

temperature: measure of how hot or cold something is, 2, 44

threads: the high points on a screw; ridges, 200

vector [VEK-tur]: an arrow that shows the strength and direction of a force, 128

volume: measure of the amount of space an object takes up, 2, 146

water vapor: water in the gas state, 44

wedge: a sloping solid that comes to a point; a simple machine, 110, 144

weight: measure of the pull of gravity on an object, 2, 118

wheel and axle: a wheel with a rod attached to its center; a simple machine, 170, 212